Chemistry
Reading and Writing the Book of Nature

Chemistry
Reading and Writing the Book of Nature

Vincenzo Balzani
University of Bologna, Bologna, Italy
Email: vincenzo.balzani@unibo.it

Margherita Venturi
University of Bologna, Bologna, Italy
Email: margherita.venturi@unibo.it

Translation by

Nick Serpone
Concordia University, Montreal, Canada
University of Pavia, Pavia, Italy
Email: nick.serpone@unipv.it

THE QUEEN'S AWARDS
FOR ENTERPRISE:
INTERNATIONAL TRADE
2013

Print ISBN: 978-1-78262-002-0

A catalogue record for this book is available from the British Library

© The Royal Society of Chemistry 2014

Original title: Chimica. Leggere e Scrivere il Libro della Natura
© 2012 Scienza Express

All rights reserved

Apart from fair dealing for the purposes of research for non-commercial purposes or for private study, criticism or review, as permitted under the Copyright, Designs and Patents Act 1988 and the Copyright and Related Rights Regulations 2003, this publication may not be reproduced, stored or transmitted, in any form or by any means, without the prior permission in writing of The Royal Society of Chemistry or the copyright owner, or in the case of reproduction in accordance with the terms of licences issued by the Copyright Licensing Agency in the UK, or in accordance with the terms of the licences issued by the appropriate Reproduction Rights Organization outside the UK. Enquiries concerning reproduction outside the terms stated here should be sent to The Royal Society of Chemistry at the address printed on this page.

The RSC is not responsible for individual opinions expressed in this work.

Published by The Royal Society of Chemistry,
Thomas Graham House, Science Park, Milton Road,
Cambridge CB4 0WF, UK

Registered Charity Number 207890

Visit our website at www.rsc.org/books

Printed and bound in Great Britain by CPI Group (UK) Ltd, Croydon, CR0 4YY

Foreword

It was the mid-1950s when I started high school (grade 10). In the chemistry class, the teacher came in, handed out a few sheets of paper (Xerox didn't exist then) and prompted us to *memorize* the Periodic Table and the valences of the elements—no reasons were given. It may be all right to memorize a poem, but to memorize science? I didn't think so. Fortunately, in the last two years of high school, I had very good teachers who explained what science was all about. It was fun. It was interesting, and there was no need to memorize anything. On completing high school, I thought I knew all the chemistry there was to know. How disillusioned I was when I began my university freshman year to discover that, in fact, I knew very little about chemistry. My class was most fortunate to have a chemistry professor who explained the notions and concepts of chemistry. We were encouraged to understand the concepts first and foremost and then learn—not memorize—learn! In my teens I wanted to become a theoretical physicist, but then found chemistry more fascinating. And so chemistry became my life's career after college and graduate school, and I have been a chemistry professor and researcher for some 40 + years.

The book, written by Balzani and Venturi, *Chemistry: Reading and Writing the Book of Nature*, is written in a way that incites the reader to understand the concepts and notions of this branch of

---

Chemistry: Reading and Writing the Book of Nature
By Vincenzo Balzani and Margherita Venturi
Translation by Nick Serpone
© The Royal Society of Chemistry 2014
Published by the Royal Society of Chemistry, www.rsc.org

science starting from atoms and molecules, chemical bonds, chemical reactions *etc.* Without the rigor of college chemistry, the authors have succeeded in explaining these concepts in very simple terms using the language of chemistry, and most interestingly make the unusual association with the language of letters and words.

The book is divided into three Parts: Part One takes the reader to a fascinating voyage into the world of atoms and molecules; Part Two discusses the chemistry of yesterday, today and tomorrow; and in Part Three the authors present approaches on how to teach science in general, and chemistry in particular.

In the last chapter of the book, the authors emphasize the development of science, its features and limitations, the role that science can play in this otherwise fragile world and, not least, the responsibility of the scientist.

Nick Serpone
Professor Emeritus, Concordia University, Montreal, Canada
Visiting Professor, University of Pavia, Pavia, Italy
Fellow of the European Academy of Sciences

Preface

This book was written to introduce young students at the junior and senior high school levels, as well as the general reader, to some of the fundamental concepts of chemistry, which is an important branch of science. Indeed, chemistry allows us to understand the many things that Nature provides in her book, which the chemist reads and considers adding chapters to; many of these have yet to be written.

Many believe that chemistry is a sterile and boring science. These views are often arrived at without justification. It is precisely such preconceived views that need to be set straight. As such, the purpose of this book is to demystify chemistry by describing, in an understandable, albeit somewhat unorthodox way, some of the fascinating notions of chemistry: atoms and molecules, chemical reactions, drugs, energy, the environment and such other topics that make chemistry a major player in daily activities. The book also addresses such other things as curiosity, creativity, fascination, poetry, beauty and ethics in science.

Perhaps, through past unfortunate experiences during their school years, some people became convinced that, compared with other subjects taught in schools, chemistry was a difficult and complex subject that was to be memorized. Quite the contrary! The concepts of chemistry are to be understood first and

---

Chemistry: Reading and Writing the Book of Nature
By Vincenzo Balzani and Margherita Venturi
Translation by Nick Serpone
© The Royal Society of Chemistry 2014
Published by the Royal Society of Chemistry, www.rsc.org

then learned. Accordingly, the book presents arguments and suggestions to be considered when teaching chemistry in secondary schools, together with a simple teaching approach, so that students can understand and come to appreciate the language of chemistry and its experimental practices. Chemistry is, after all, an experimental science!

We would like to thank the members of our research group for daily discussions on science and society issues. In particular we are grateful to Alberto Juris who has also contributed to improving several of the figures.

Vincenzo Balzani
Margherita Venturi
University of Bologna, Bologna, Italy

Contents

Part One
A Fascinating Voyage: The World of Atoms and Molecules

Chapter 1
Chemistry: An Essential Science   3

1.1   Education, Science, Culture   3
1.2   Chemistry's Image   5
1.3   The Importance of Chemistry   6
1.4   The Relevance of Chemistry Among the Sciences   7

Chapter 2
Atoms and Molecules: The Language of Chemistry   9

2.1   The Perfume of a Rose   9
2.2   What is the World Made of?   10
2.3   Comparing Matter with Language   15
2.4   A Bit of History   18
     2.4.1   From the Greek philosophers to Lucretius   19
     2.4.2   From Dalton to Cannizzaro   20
     2.4.3   The Last 150 Years   22

Chemistry: Reading and Writing the Book of Nature
By Vincenzo Balzani and Margherita Venturi
Translation by Nick Serpone
© The Royal Society of Chemistry 2014
Published by the Royal Society of Chemistry, www.rsc.org

Chapter 3
The World of Molecules — 24

3.1 The Chemical Bond — 24
3.2 The Dimensions of Molecules — 27
3.3 The Names of Molecules — 29
3.4 The Formulas of Molecules — 30
3.5 Molecular Models — 34
3.6 Attention to Details! — 37

Chapter 4
Chemistry in Action: The Reactions — 39

4.1 Transformation of Chemical Species — 39
4.2 Oxidation–Reduction (redox) Reactions — 42
4.3 Acid–Base Reactions — 44
4.4 Chemical Equations — 46
4.5 Why Reactions Occur — 48
4.6 Reactions and Time — 50

Chapter 5
Beyond Molecules: From Chemistry to Biology — 53

5.1 From Molecules to Supramolecular Systems — 53
5.2 From Supramolecular Systems to Cells — 57
5.3 From Cells to Man — 61
5.4 Genetic Engineering — 62
5.5 Beyond the Scale of Complexity — 63

Part Two
Chemistry: Yesterday, Today and Tomorrow

Chapter 6
Reading and Writing with Molecules — 67

6.1 The Chemist: Explorer and Inventor — 67
6.2 The Merry-Go-Round of Curiosity — 68
6.3 Natural Molecules — 71
6.4 How We Got to *Aspirin* — 72
6.5 Artificial Molecules — 74

Chapter 7
Creativity and Beauty — 78

- 7.1 For Better or for Worse — 78
- 7.2 Beautiful Molecules — 80
- 7.3 Chemistry in the Words of Scientists and Writers — 81
- 7.4 Intelligent Molecules — 84
- 7.5 Molecular Machines — 88

Part Three
Teaching and Science

Chapter 8
On Teaching: What and How — 93

- 8.1 Chemistry as a School Subject — 93
- 8.2 What to Teach — 94
  - 8.2.1 First Thematic Nucleus: Atoms, Molecules, Ions and the Chemical Bond — 95
  - 8.2.2 Second Thematic Nucleus: The States of Aggregation of Matter and the Solutions — 96
  - 8.2.3 Third Thematic Nucleus: The Chemical Reactions — 97
  - 8.2.4 Fourth Thematic Nucleus: Chemistry in Everyday Life — 100
  - 8.2.5 Fifth Thematic Nucleus: Chemistry Toward the Future — 100
- 8.3 How to Teach — 101

Chapter 9
Today's Science: Objectives, Implications and Limits — 106

- 9.1 The Development of Science — 106
- 9.2 Will Science Come to an End? — 107
- 9.3 Characteristics and Limitations of Science — 109
- 9.4 The Role of Science in a Fragile World — 111
- 9.5 The Social Responsibility of the Scientist — 112

Subject Index — 114

Part One
A Fascinating Voyage: The World of Atoms and Molecules

Part One

CHAPTER 1

Chemistry: An Essential Science

1.1 EDUCATION, SCIENCE, CULTURE

If you think education is expensive, try ignorance. This succinct phrase, said by Derek Bok, former President of Harvard University (USA), has been cited many times in intellectual and political circles, particularly during times of repeated cuts to education, culture and scientific research.

> If you think education is expensive, try ignorance

Evidently, the statement was meant to provoke a reaction, and it did. Whenever a country leaves its people—particularly its young people—in ignorance, the cost to educate them will be far greater tomorrow than it would be today. Far too many countries pay scarce attention to education, providing it with no resources, financial or otherwise, particularly in the scientific and technological fields. It is these fields that will provide their population with an understanding of today's issues and prepare their young people to build a better future. Countries must recognize that science and technology are the fundamental pillars of their economic wellbeing.

Chemistry: Reading and Writing the Book of Nature
By Vincenzo Balzani and Margherita Venturi
Translation by Nick Serpone
© The Royal Society of Chemistry 2014
Published by the Royal Society of Chemistry, www.rsc.org

Chemistry is an important component of the scientific establishment and technological development of any country. Accordingly, it should be an important part of the cultural heritage of its people.

Unfortunately, through the years chemistry has gained an undeserving reputation as a tedious and difficult school subject. Furthermore, it is often represented in a frightening manner in printed and electronic media. It is not surprising, therefore, that many intellectuals, who may have been confronted with an adverse short-lived encounter with chemistry during their school years and conditioned by a misinformed media, make gratuitous reckless judgments about chemistry. They often refer to this scientific field with such derogatory statements as:

I have never understood chemistry but live well just the same.

Everything that is chemical is bad for one's health.

Chemistry is a waste of time and causes pollution.

The chemical industry

Perhaps few recognize that many of the things we use daily are produced by the chemical industry, which is an important manufacturing sector in developed countries. The sector transforms raw materials (for example, petroleum, natural gas, air, water and minerals) into no less than 70 000 different products, making the chemical industry one of the largest employers, which makes a significant contribution to a country's Gross Domestic Product (GDP). For instance, in the USA, the chemical industry employs directly more than one million workers, and is responsible for more than 2% of the USA's GDP.

Rarely does one reflect on the fact that it is the chemists, after all, who develop and produce the pharmaceutical products that we often take for granted—medicines that we have no problem taking when sick. And yet, few people appreciate the fact that chemistry is a field of science that can, through continuous research and development (R&D), resolve some of the more important problems that plague society. And few seem to realize that chemistry, in addition to being useful and important, is also a fascinating field of science. Lacking the basic fundamentals of chemistry, which are typically taught at High School level these days, is indeed most regrettable. It is as regrettable as having never read Dante's Divine Comedy, or one of Shakespeare's plays or the interesting writings of some more recent authors—for example, James Joyce, Ernest Hemingway or Mark Twain.

1.2 CHEMISTRY'S IMAGE

Unquestionably, chemistry is too often viewed negatively. Giorgio Nebbia, an authoritative chemist and a member of Italy's Senate, often asked himself why the word *chemistry* is such a frightful word. Whenever one brings up a chemical argument in a conversation, one notices a sudden silence and drop of interest. And when people subsequently discover that the person making the argument is a chemistry professor, there is a sudden manifestation of surprise—if not outright shock. Why would an intelligent, normal person want to teach a subject as incomprehensible and as hazardous as chemistry?

Somewhat teasingly and somewhat seriously, the prestigious American *Journal of Chemical Education* published a one-minute Chemistry course to help chemists explain, in a simple and clear manner during, for example, an elevator ride or other short encounter, what chemistry is all about and why it is an important branch of science. Later, the *American Chemical Society* introduced a course for would-be *Ambassadors of Chemistry* to teach chemists how to dialog with the public.

Chemistry seems to make the news in the printed and electronic media *only* when ecological disasters and related calamities occur; the dominant message being that, in addition to being a boring and difficult school subject, chemistry is responsible for such disasters. It is something to be suspicious of. In fact, chemistry is seen as something that should be avoided at all costs.

> **Whoever said that *chemical* should always rhyme with *immoral*?**

An advertisement for prosciutto, which is produced in Parma, appeared a few years ago in many Italian newspapers, asserting in large letters: *Purezza sì Chimica no* (*Purity yes Chemistry no*). Yet people fail to realize that everything in those prosciuttos is chemical in nature, including its distinctive purity and its appeal to the keenest and most sophisticated palate. It is no surprise then that, from advertisements such as these and the many similar asinine statements by the printed and electronic media, the common man or woman believes that the word *chemical* is the opposite of *natural* and hence is synonymous with *artificial*.

But whoever said that *artificial* must necessarily be the same as *immoral*? Chemophobia is so entrenched in people's minds today, that a few years ago the Chemical Industry Association found it necessary to buy space in a number of major daily newspapers to publicize the role and benefits of chemistry.

1.3 THE IMPORTANCE OF CHEMISTRY

It is absolutely necessary to explain to people—in particular, young people—that everything around us and within us, in fact, involves chemistry. Chemistry includes many of the indispensable natural phenomena of life; a clear example is the process known as *photosynthesis*. The so-called *artificial products* of basic importance to society—for instance, drugs, fertilizers, plastics, semiconductors and detergents (among others)—are all the result of chemical research. *We are, after all, made of chemistry!* Our bodily functions are nothing more than an ensemble of chemical reactions

> **Chemistry is around us and within us**

Life is chemistry in action! Without being pretentious, we can easily state that all life manifestations, including those that we can classify as mental—learning, memory, thought, experience, dreams—are, in the final analysis, the result of chemical reactions, some of which are so complex that we are still unable to interpret them. Perhaps the situation will change as our knowledge advances. The history of science tells us that what seems impossible today could very well be possible or otherwise be self-evident tomorrow.

We also need to point out that chemistry should not be confused with the harmful effects it can potentially have when chemical products are used improperly. For instance, a life can be saved with certain chemical products (drugs, medicines); a life can be taken away with certain other chemical products (poisons). However, chemistry cannot be faulted for the improper use of chemical products. Just as a knife can be used to cut bread and to kill a human being, it is not the fault of the knife—it is the fault of whoever uses it inappropriately!

1.4 THE RELEVANCE OF CHEMISTRY AMONG THE SCIENCES

Increasingly, chemistry will continue to shape our future because it is a scientific discipline at the center of many others (Figure 1.1). With its language of molecules, chemistry invades and pervades numerous other fields of knowledge, acting, as it were, as a bridge between them.

> The language of Chemistry pervades several fields of knowledge

Chemistry has given new perspectives to biology, which—in its more advanced version—has come to be known as molecular biology. In turn, the latter has profoundly revolutionized the field of medicine. In the near future, the problems of medicine, and perhaps even thoughts, feelings, and emotions will be discussed in molecular terms—that is, in chemical terms. Chemistry is also the very

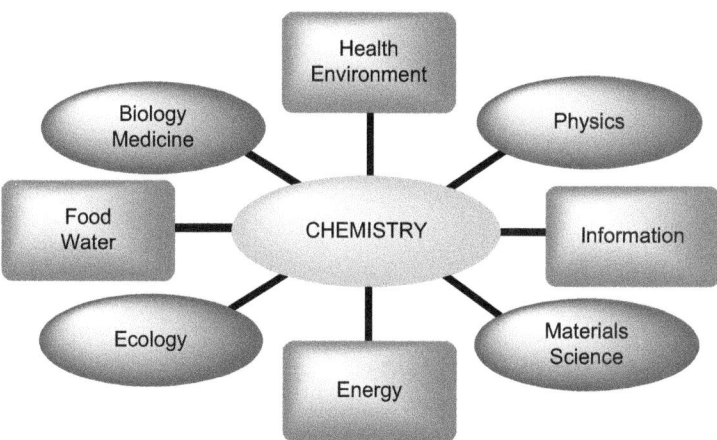

Figure 1.1 The importance of chemistry.

foundation of other emerging disciplines: for example, materials science and ecology.

In fact, chemistry is at the base of all the benefits that, in some unconscious way, we enjoy daily. As a simple example, *textiles* are used to make clothes that protect us from the cold and rain.

In one form or another, chemistry will impact our future for a long time to come because only through chemistry will we be able to resolve the four great problems that mankind faces and that we need to tackle to safeguard our planet Earth (Figure 1.1): food and water; health and environment; energy; and information.

Chemistry will certainly make significant contributions to these issues as the chemist—in his work—is a visionary, is humble and respects the world around him. These assets are necessary so that scientific and technological activities remain truly human.

CHAPTER 2

Atoms and Molecules: The Language of Chemistry

2.1 THE PERFUME OF A ROSE

The things we all admire, such as stars in the night sky or flowers in a field, are even more admirable when our mind succeeds in penetrating the intimate soul of Nature through science.

Let's consider, for example, the fragrance of a flower. The flower scatters small entities in the air, which are invisible even under a microscope because of their small dimensions; they are smaller than a billionth of meter, or a million times smaller than the thickness of a hair. Even though they are small, these entities that the chemist calls *molecules* possess specific shapes and properties. For instance, the molecules released into the air by a rose are significantly different from those released by a cyclamen (Figure 2.1). When the molecules released by the rose reach the nostrils, they encounter nasal receptors in the mucous cavities. These receptors consist of molecules that possess appropriate shapes and properties, such that they can recognize the molecules of the rose by combining with them in a manner reminiscent of a lock and key. As a result of this combination, the nasal receptors signal to the brain through the nerve endings of our organism,

---

Chemistry: Reading and Writing the Book of Nature
By Vincenzo Balzani and Margherita Venturi
Translation by Nick Serpone
© The Royal Society of Chemistry 2014
Published by the Royal Society of Chemistry, www.rsc.org

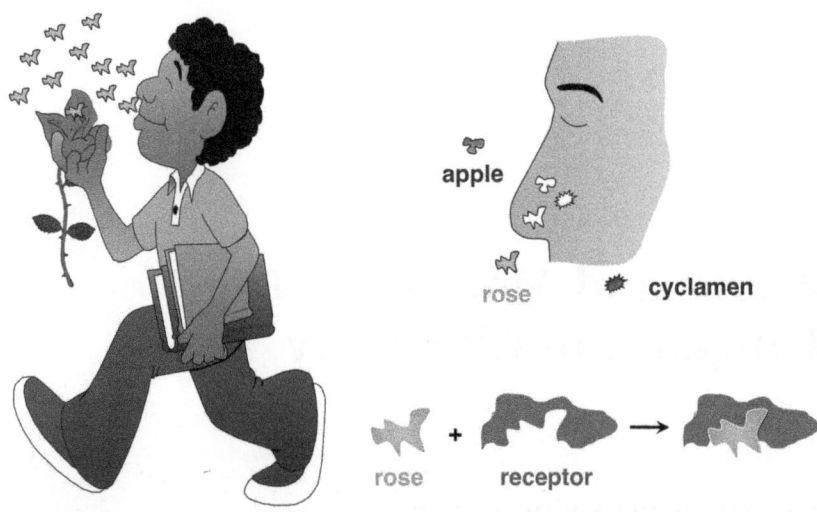

Figure 2.1 Each scent is due to a specific type of molecule.

and provoke that pleasant feeling we have come to identify as *the perfume of a rose*.

By contrast, the molecules released by the cyclamen possess different shapes and properties. When these molecules interact with various nasal receptors, diverse from those that recognize molecules of the rose, they generate a different nerve impulse such that our brain now recognizes the scent as *the perfume of a cyclamen*.

These two simple examples suffice to emphasize that having no knowledge of the world of molecules is not simply a deficiency of scientific culture, but also limits the ability to capture emotions and understand the complexity and beauty of Nature.

2.2 WHAT IS THE WORLD MADE OF?

Even though the concepts of the *element*, the *atom*, and the *molecule* are now rather common in their simplest form, it is nonetheless essential that we clarify their different meanings.

The elements

One or more fundamental constituents can be found in all substances that exist in Nature, or that have otherwise been produced by man. These are the so-called *elements*. There are approximately 100 such elements. In many cases, their names may be familiar to you; for instance, hydrogen, oxygen, nitrogen, carbon, sodium, potassium, iron (and so on).

Every element possesses properties that are different from those of other elements. Elements are typically identified by means of their first letter, or else by the first two letters of their name—often the Latin name—that was attributed to them at the moment of their discovery; for example, H for hydrogen, O for oxygen, N for nitrogen, C for carbon, Na for sodium (from the Latin word, *Natrium*), K for potassium (from *Kalium*), Fe for iron (from *Ferrum*), etc.

In the universe

- The first element formed was hydrogen.
- The most abundant elements are hydrogen and helium.

On Earth

- The most abundant elements, in terms of their percentage by weight, are: oxygen (48.9%), silicon (26.3%), aluminum (7.7%), iron (4.7%) and calcium (3.4%).
- The rarest (non-radioactive) element is krypton (Kr), present by weight percentage equal to 1.9×10^{-8}% (0.000 000 019%).

In the human body

- The most abundant elements in terms of their percentage by weight are: oxygen (65.4%), carbon (18.1%), hydrogen (10.1%), nitrogen (3.0%), calcium (1.5%), phosphorous (1.0%) and sulfur (0.25%); this means that in a 70 kg man there are approximately 45.5 kg of oxygen, 12.6 kg of carbon, 7.0 kg of hydrogen, 2.1 kg of nitrogen, 1.1 kg of calcium, 0.7 kg of phosphorous and 0.2 kg of sulfur.
- The least abundant elements, which are nonetheless essential for sustaining life, are chromium (Cr), cobalt (Co), and molybdenum (Mo), each of which is present from 3 to 5 mg.

Based on their properties, the elements have been ordered in a table called the Periodic Table (Figure 2.2), sometimes also

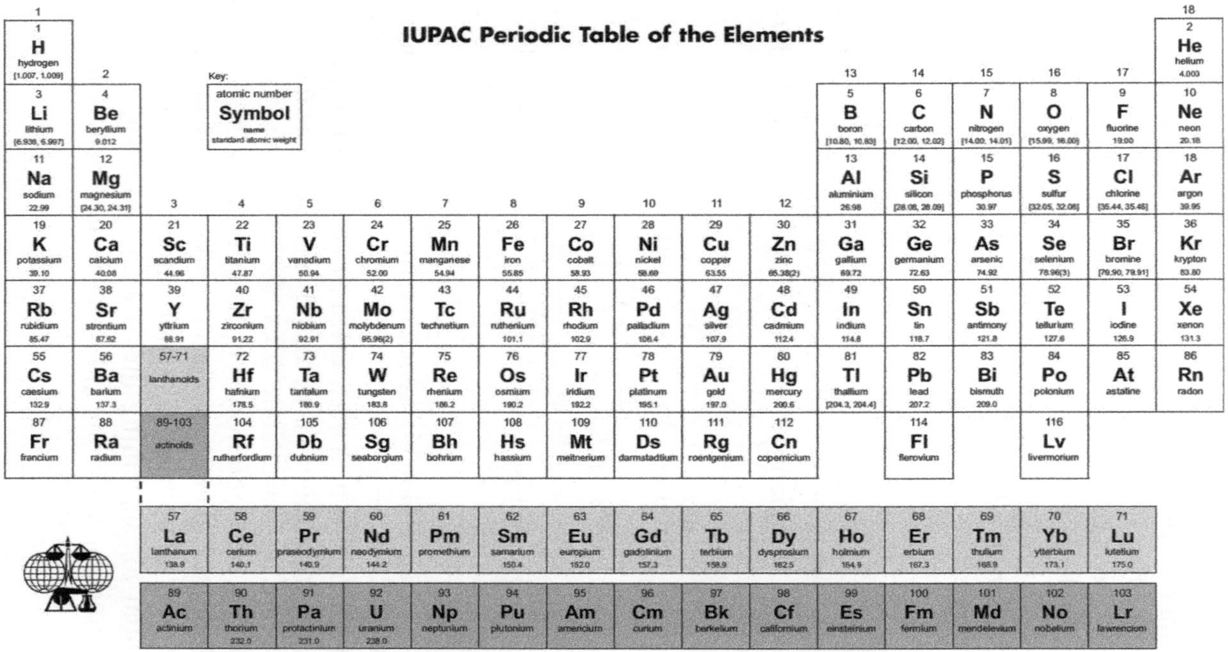

Figure 2.2 The current and commonly used Periodic Table: Chemistry's icon. Copyright 2005 IUPAC—The International Union of Pure & Applied Chemistry.

Atoms and Molecules: The Language of Chemistry

Figure 2.3 The original Periodic Table as proposed by Mendeleev.

referred to as the Periodic System, a term that the Italian chemist and writer Primo Levi used as the title of one of his famous books.

The birth of the Periodic Table occurred in 1869, from the work of the Russian chemist Dmitri I. Mendeleev (Figure 2.3) who, without understanding the reasons why, was the first to recognize some of the existing similarities between the properties of the elements.

According to many scientists, Mendeleev's scheme was one of the most brilliant coups de force of the last ten centuries. For decades, the Periodic Table has been treated as nothing short of magical.

| Chemistry's icon |

Even though the reasons for the similarities between various elements are now well known, the Periodic Table retains all its fascination. We can catch a glimpse of the intrinsic and profound order in Nature because of the clear order of the elements. The Periodic Table summarizes a good deal of chemistry in a concise and unique manner. No other scientific discipline can boast a similar achievement.

| The atoms |

The smallest particle of an element is its *atom* (from the Greek word: *atomos*, meaning not divisible). In fact, we now know that the structure of an atom is far more complex than first thought. Without entering into any detail at this point, suffice it to note that the chemical characteristics of an atom are fundamentally associated with the number of electrons it contains. The atoms of the same element—for example, those that constitute a piece of pure gold—all have the same number of electrons and so have the same chemical properties. Atoms of different elements—for example, atoms of gold and iron—have different properties because they contain a different number of electrons.

Atoms are particles that are typically spherical in shape and different in size depending on the element. As an example, the radius of a carbon atom is approximately 0.000 000 000 08 meters, whereas the radius of an iron atom is 0.000 000 000 12 meters. To express the dimensions of such small objects, scientists have adopted a unit of measure known as the *nanometer* (abbreviated as nm), which represents a billionth of a meter. In other words, 1 nm equals 10^{-9} meter, or 0.000 000 001 meter. The radius of the carbon atom is therefore said to be 0.08 nm and that of the iron atom to be 0.12 nm.

To understand how small the atoms are, it is useful to consider a few examples to help you visualize them. The tip of a pencil is made of graphite, a solid that consists of carbon atoms. If you were to trace a line 3 centimeters (cm) long and 0.2 millimeters (mm) thick with a pencil on a sheet of paper, more or less like this: _____, a maxi-army of carbon atoms would be left on the sheet: approximately one million rows aligned one close to the next, each consisting of approximately one hundred million of these invisible little soldiers.

If you are still not overwhelmed, then note that, if the atom had the dimensions of a point (or a period, "."), the height of a man would be more than a thousand kilometers. In addition, you might also be surprised to know that every component of matter that is hardly visible—for example, a grain of sand—contains more atoms than the number of stars present in our Milky Way galaxy.

The molecules

In general, atoms are not isolated species, but tend to combine, ultimately giving rise to molecules. Indeed, molecules can be constituted either from the same atoms (molecules of the elements) or from different atoms to produce molecules of compounds. The number of atoms of the same type in a molecule is indicated by a subscripted number.

Atoms and Molecules: The Language of Chemistry

For instance, the molecule of the element oxygen, formed simply by combining two atoms of oxygen, is represented as O_2, while that of the compound water, which consists of two atoms of hydrogen and one atom of oxygen, is indicated as H_2O. Other molecules are constituted from a greater number of atoms. For example, a molecule of glucose consists of a total of 24 atoms: 6 atoms of carbon, 12 atoms of hydrogen and 6 atoms of oxygen. Accordingly, glucose is represented as $C_6H_{12}O_6$. As we shall see later, however, there are molecules that are far more complex.

Because molecules are created from a small (or relatively small) number of atoms, their sizes are also in the billionths of a meter (that is, in the nanometer) scale. Molecules can thus be considered as the invisible bricks that make up the world around us. To enter into this infinitely small dimension is to discover how molecules are made and how they react. This will surely be a most fascinating voyage that we will be taking soon.

2.3 COMPARING MATTER WITH LANGUAGE

To better understand the basic nature of atoms and molecules, as well as the complexity of the real world around us, it will prove instructive to compare the world of *matter* with the world of *language* (Figure 2.4).

To some extent, the insightful analogy between the world of *words* and the world of *things* had been guessed and described more than 2000 years ago by the Roman poet and philosopher

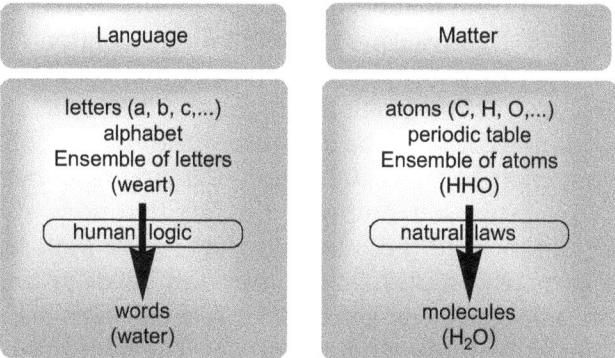

Figure 2.4 A comparison between the structure of language and the structure of matter is very useful in understanding the complexity of the material world.

Titus Lucretius Carus—better known as Lucretius—who stated in his philosophical poem *De rerum natura*:

> *in fact they are always the same [letters] to indicate the sky, the sea, the lands ... but their different order distinguishes the names of things. The same happens in bodies: as soon as the encounters vary, the motions, the order, the position, the shapes of matter, so do the bodies themselves have to change.*

The words of matter

Every language is based on elementary graphical units that we call letters (Figure 2.4). In the English language there are 26 letters (a, b, c, *etc.*) that make up the English alphabet. Similarly, the elementary units of matter are the atoms, approximately 100 of them—they are listed in the Periodic Table (Figure 2.2). The analogy between language and matter gets even more interesting if we consider (as seen earlier) that the atoms of the different elements are represented by letters (for example, H for hydrogen, O for oxygen, C for carbon, *etc.*). Everything in a language is made up of letters, so too is all matter made up of atoms.

In a language, letters of the alphabet are generally not used in isolation, but are arranged in groups, following a logic invented by man. These groups of letters form words. As an example, arranging the letters a, e, r, t and w in an appropriate fashion yields the word *water*. The same takes place with matter from the moment that, in the reality of matter, rather than finding isolated atoms, we commonly find their combinations formed according to rules imposed by the laws of Nature. These combinations of atoms are the molecules. That is, the molecules represent the combinations of atoms, just like words result from the combination of letters. Accordingly, molecules can be considered the "words" of matter.

The connection between the letters of a language that constitute a word is expressed simply by the fact that they are written (and said) one next to the other. In the case of molecules, the situation is more complex. The connection between two atoms present in a molecule is generally indicated with a dash (–), which joins the symbols of the two atoms. The molecule of water, which is often indicated with the formula H_2O to point out that it is made of two atoms of hydrogen and one atom of oxygen, is more appropriately represented by the formula H–O–H.

The number of combinations of atoms to form molecules, and letters to form words, is practically infinite. In reality, however, not all the combinations have meaning, because to have meaning the combinations must obey some well-defined rules. In the case of the letters a, e, r, t and w, the word *water*, for example, is an appropriate combination of letters, whereas the word *weart* is a meaningless combination. Here, the terms *appropriate* and *meaningless* are judged solely on the basis of conventions that have been established in formulating the language. Even in the case of atoms, we may encounter appropriate and meaningless combinations. For example, H–O–H is an appropriate combination, whereas O–H–H is a meaningless combination. In the case of atoms, however, appropriate and meaningless simply mean that the combination H–O–H exists in the reality of matter, whereas O–H–H does not. This is due to the laws of Nature and to the intrinsic properties of the atoms. Within this context, the hydrogen atom, H, cannot be connected to two other atoms. In other words, H cannot occupy the mid-position of the combination. Not only does the combination O–H–H not exist in Nature, but it is also not possible to put together these three atoms in such a fashion, not even artificially in a chemical laboratory.

Every word is constituted by an aggregate of letters with its own structure, in the sense that the members (the letters) are in an established relationship that gives a unique and specific meaning to the aggregate. Similarly, a molecule is an aggregate of atoms that has its own structure. The relationship between the atoms (relative positions and interactions) imparts unique and specific properties to the aggregate.

The most aromatic molecules are:

Grapefruit flavor	Cork flavor	Wine bouquet

> Our taste can perceive the presence of a compound that imparts the flavor of grapefruit when only 2 mg are dissolved in one hundred million liters of water. For the compound responsible for the flavor of "cork" in a bottle of wine, it is sufficient for only 10^{-9} g be present for us to be disgusted with the wine. In addition, our sense of smell is so sensitive to the compound that confers certain wines a sweet mixed bouquet of coconut and resin in order to perceive its presence when only 10^{-14} g are present in a liter of air.

A *word* is more than just the letters that constitute it. Likewise, a molecule is much more than the atoms from which it is formed. By themselves, the components cannot determine completely the properties of the combinations.

Molecules, therefore, are the words of chemistry, the words of matter, the words of things around us. Just as there exist short words (that is, made up of a small number of letters) and long words, so there exist molecules consisting of a small and a large number of atoms. For instance, the molecule of carbon dioxide, commonly also known as carbonic anhydride, CO_2, is constituted by two atoms of oxygen and one of carbon. Larger molecules are composed of a larger number of atoms such as, for example, the molecule of ethanol, C_2H_6O, formed by 2 carbon atoms, 6 hydrogen atoms, and 1 oxygen atom. But while some words rarely contain more than 10 to 15 letters (the longest word in the Italian language, *precipitevolissimevolmente*—meaning *precipitously*, or as fast as possible—is made up of 26 letters; the longest word in Shakespeare's writings, *honorificabilitudinitatibus*, is made up of 27 letters), molecules can also be made up of hundreds or thousands of atoms. For example, hemoglobin consists of 9072 atoms; its molecular formula is $C_{2954}H_{4516}N_{780}O_{806}S_{12}Fe_4$.

2.4 A BIT OF HISTORY

Before venturing into the world of molecules, it is useful to journey again over the principal and laborious stages that have allowed scientists to demonstrate the existence of the fundamental constituents of real matter.

The concept of a *molecule* has had a difficult time asserting itself in the history of science, because it has often been confused by philosophers first, and by scientists later, with the concepts *atom* and *element*. The word "molecule" is a term derived from the Latin word *molecula*, a diminutive of *moles* (mass). However, as we shall see later, its meaning today is much richer and more complex than could have been envisaged from its simple etymological derivation.

2.4.1 From the Greek philosophers to Lucretius

The notion that Nature's complexity can be reduced to a small number of elementary substances was developed long ago. According to Empedocles (a Greek pre-Socratic philosopher and a citizen of Agrigentum, a Greek city in Sicily; *ca.* 483–423 B.C.) the fundamental elements were: air, fire, water and earth. The concept that matter is discontinuous—that is, it is not sub-divisible into an infinite number of smaller parts—goes back to the Ionian philosopher Leucippus (second half of the 5th century BC). The idea that all things are constituted of an aggregation of indivisible particles, namely atoms, was developed subsequently by the Greek philosopher Democritus (*ca.* 460–370 BC), a student of Leucippus, and then elaborated again by the Greek philosopher Epicurus (341–270 BC). According to the latter, matter, of which man is also constituted, is formed of atoms that are in continual movement and come in various shapes. By interacting with each other, the atoms can join in various ways, thereby giving origin to all possible shapes of matter.

In his philosophical poem *De rerum natura*, the Latin poet Lucretius (*ca.* 98–55 BC) exposed the atomistic theory of Epicurus on the nature of man and the world in some detail. Rediscovered during the Renaissance period (1400 to the beginning of the 1600s), this poem was strongly influential on the emergence of scientific thought. The conviction that matter was really made up of a corpuscular nature was asserted in the 16th century, contrary to what had been commonly believed for centuries from the ideas of the Greek philosopher Aristotle.

2.4.2 From Dalton to Cannizzaro

It was only in 1807 that the atomic hypothesis was placed on a scientific basis by the English scientist John Dalton (1766–1844) who gave it a very precise chemical meaning:

(1) all materials are constituted of indivisible and non-transformable atoms;
(2) there are as many different types of atoms as there are chemical elements;
(3) compounds are obtained through a combination of atoms of various elements according to well-defined numerical relationships.

Using little spheres, dashes, asterisks, grids and others, Dalton indicated how various atoms and their compounds could be subdivided into binary, ternary, *etc.* according to how the *primary atoms*, as he called them, became a part of the compounds (Figure 2.5).

It took nearly 100 years of experimental research and thoughtful consideration for the atomic theory to be accepted *in toto*, and for the various types of atoms to be characterized satisfactorily and ordered according to Mendeleev's Periodic Table.

In parallel with the many discussions on atomic theory, another issue regarded the smallest entity a substance could have so as to be in a free state, and to possess the properties of that same substance. This entity was called various names: a particle, an element, an integrated molecule, or simply a molecule. During the same period in which Dalton pursued his studies on the atomic hypothesis, the French scientist Gay-Lussac (1778–1850) showed, by means of a series of experiments between gaseous substances, that in equal volumes of a gas, under identical conditions of temperature and pressure, the same number of particles is contained, or a simple multiple thereof. Even before Dalton had formulated his atomic theory, another French scientist, Lavoisier (1743–1794), in 1781 had discovered that water, until then considered an element, was in fact a compound formed from two simpler substances, namely hydrogen and oxygen.

Atoms and Molecules: The Language of Chemistry

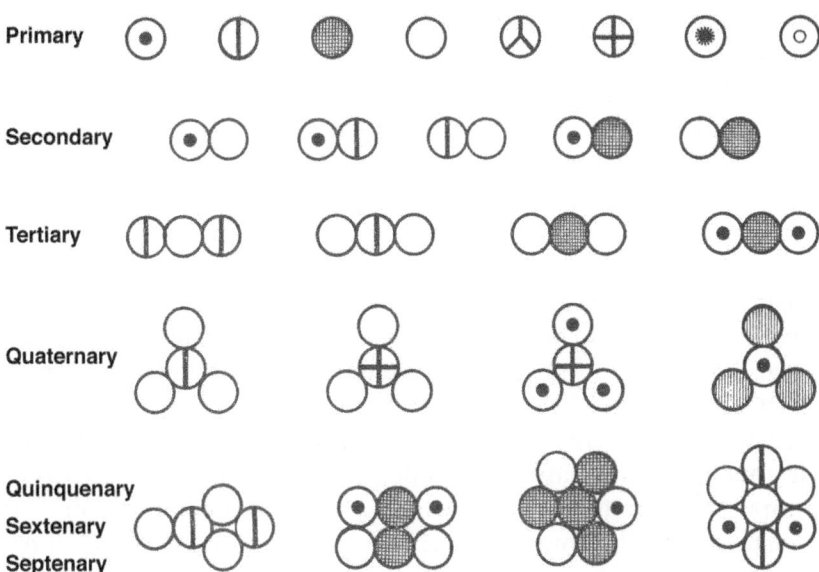

Figure 2.5 Symbolism used by Dalton to represent atoms. Adapted from http://oi29.tinypic.com/e80qjs.jpg; see also http://tinypic.com/ view.php?pic = e80qjs&s = 3#.UvjTT_ldXX4.

Dalton subsequently proposed that the smallest quantity of water to exist in the free state and to possess the property of the same substance (that is, the water molecule) was the combination of a hydrogen atom and an oxygen atom. Using the chemical symbols introduced in the meantime by the Swedish chemist Berzelius (1779–1848), he represented the water molecule simply as O–H. This representation of water was, however, in contradiction with the results found by Gay-Lussac, in that two volumes of hydrogen react with one volume of oxygen to produce two volumes of water vapor.

In 1811, the Italian chemist Amedeo Avogadro (1776–1856) realized that this contradiction could be resolved by assuming that, in an equal volume of gas there is an equal number of molecules and not atoms, since a molecule can be constituted by many atoms. To explain the results of Gay-Lussac by Dalton's atomic theory, it was therefore sufficient to recognize that the oxygen and hydrogen molecules were formed from two atoms each, H_2 and O_2, and that the water molecule was formed from two atoms of hydrogen and one of oxygen: H_2O. Avogadro's idea, however, was rejected by a large number of chemists of his time. Berzelius, for example, who was the most influential chemist of the first half of the 19th century, continued to believe that equal volumes of gas did *not* contain the same number of molecules, but contained the same number of atoms. It was only many years later that Avogadro's hypothesis was proposed again with great success by another Italian chemist, Stanislao Cannizzaro (1826–1910).

At the first Chemistry Conference held in 1860 in Karlsruhe, Germany, Cannizzaro succeeded in convincing his contemporaries that it was necessary to distinguish the concept *molecule* from the concept *atom*. In the case of the *molecule*, they had to accept the notion that a molecule was the smallest quantity of a substance that retains its characteristics and participates in reactions. By contrast, the *atom* had to be considered as the smallest quantity of an element present in the molecule of its compounds. Thus, nearly 150 years have passed since chemists succeeded in demonstrating the existence of the invisible and mysterious entities that today we call *atoms* and *molecules*.

2.4.3 The Last 150 Years

With advances in research, especially on substances from vegetable and animal organisms, various concepts have been developed:

- formation of bonds between atoms;
- chemical affinity, which denotes the tendency that certain atoms have in joining with others;
- valence, which refers to the number of other atoms with which an atom can combine.

It is now clear that if it is true that everything is made up of atoms, then it is also true that the atoms are a very reactive species and therefore are not isolated. As a rule, atoms bond with other atoms following very precise laws, and form molecules according to the specific electronic structure of each atom. Chemists have uncovered, albeit slowly, the world of molecules by performing experiments on macroscopic quantities of matter using various experimental techniques.

CHAPTER 3

The World of Molecules

3.1 THE CHEMICAL BOND

The most important property of atoms is their capacity to *combine*—that is, to bond to other atoms to form molecules according to well-defined laws of Nature. The specificity of the chemical bond is related to the number of electrons and the manner in which they are arranged in the atom. The bond between two atoms—that is, the *glue* that keeps them together—originates from the sharing of a couple of electrons. In the majority of cases, the bonds that maintain the atoms together in a molecule, generally referred to as *covalent bonds*, are strong and only if sufficient energy is available in the form of thermal energy (heat), light, or electrical potential (among others) can such bonds be broken.

Other types of bonds are the ionic bond present in ionic crystals, for example, in sodium chloride (kitchen salt) and in calcium oxide (lye), and the metallic bond that exists, for example, in an iron bar or in a gold ingot.

| Coupling between atoms |

The number of covalent bonds that an atom can make depends on the number of electrons that can be shared with its neighboring atoms.

The World of Molecules

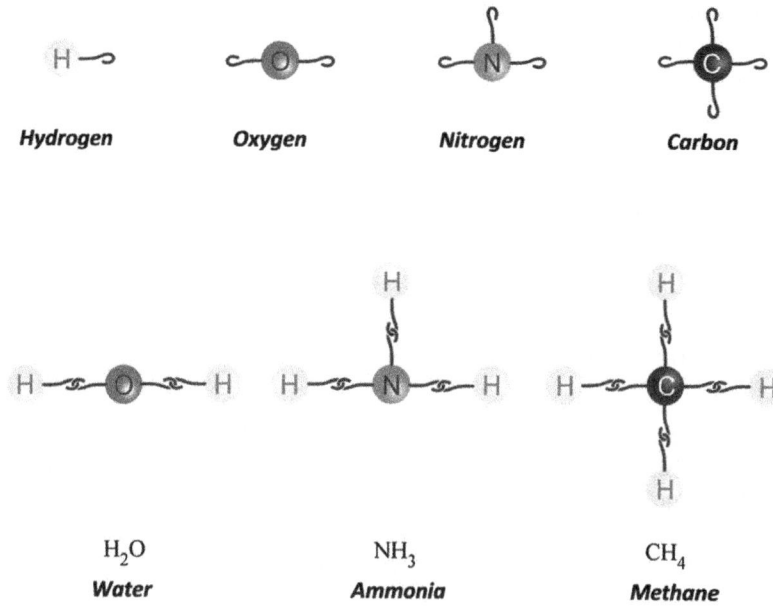

Figure 3.1 Schematic representation of bonds between atoms. Each atom has hooks with which it can couple with other atoms to give rise to molecules.

In the rather simplistic schematic representation of Figure 3.1, these electrons are depicted as little *hooks*. For instance, the hydrogen atom, H, has only one hook, while the oxygen atom, O, has two hooks, the nitrogen atom, N, has three such hooks, and the carbon atom, C, has four hooks. Hence, it is easy to understand how molecules are formed. Each atom uses its hooks to link to the hooks of the other atoms. In this way, the oxygen atom, O, with its two hooks, can couple with two hydrogen atoms, H, each of which has only one hook. Thus, this combination produces a molecule of water, H_2O.

In an analogous fashion, the nitrogen atom, N, with its three hooks, can couple with three atoms of hydrogen, H, to give the molecule of ammonia (NH_3), whereas the carbon atom, C, with its four hooks, can combine with four atoms of hydrogen to give the methane molecule, CH_4.

It is now common practice to use dashes rather than hooks—as chemists do—to represent the bonds that unite the symbols of the bonded atoms, and so the molecules of H_2O, NH_3, and CH_4, reported in Figure 3.1, can be represented simply as displayed in Figure 3.2a.

Certain atoms, such as the hydrogen atom H, form only single bonds, represented by one single dash, while others can also form double or triple bonds, represented by two or three dashes, respectively. Thus, for example, the two oxygen atoms O that constitute the oxygen molecule, O_2, are united by a double bond (O=O), whereas the two nitrogen atoms N that make up the molecule N_2 are connected by a triple bond, N≡N (Figure 3.2b). The carbon atom, C, is one of the more common atoms encountered in Nature and is present in all molecules of living organisms. It has the peculiarity of being able to form single, double and triple bonds with another carbon atom or with certain other atoms (Figure 3.2c). With regard to the strength of such bonds (referred to as *bond strength*), a single bond is weaker than a double bond, which in turn is weaker than a triple bond. Since the nitrogen molecule, N_2, consists of two nitrogen atoms

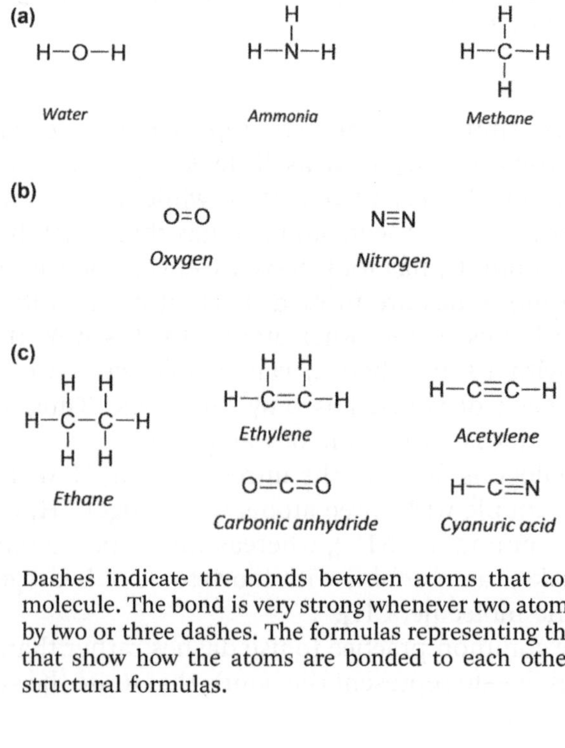

Figure 3.2 Dashes indicate the bonds between atoms that constitute the molecule. The bond is very strong whenever two atoms are joined by two or three dashes. The formulas representing the molecules that show how the atoms are bonded to each other are called structural formulas.

The World of Molecules

connected by a triple bond, the N_2 molecule is a very difficult molecule to break up as it requires a considerable amount of energy.

Needless to say, the problem of representing the bonds is of fundamental importance in interpreting and explaining the world of molecules.

3.2 THE DIMENSIONS OF MOLECULES

As we alluded to earlier, molecules are *objects* that have dimensions in the order of nanometers (nm); recall that a nm is a billionth of a meter. A molecule of water, for example, whose size is approximately 0.2 nm, is so small that a drop of water contains about 10^{21} (that is, one thousand billion billion) such molecules. There are so many, in fact, that if we could distribute them to every inhabitant of the Earth, everyone would receive about 200 billion molecules. If we counted each molecule at the rate of one per second, it would take approximately thirty thousand billion years to count them all.

Objects of such small dimensions are beyond our daily experience and common experimental investigations. In this regard, taken singularly, molecules can neither be seen, nor weighed, nor measured.

| **Chemists: blind but with sensitive fingers** |

Despite such difficulties, however, chemists have learned to distinguish the molecules, to determine their mass, to establish their atomic composition, to evaluate their dimensions, to guess at their shape, and to characterize their properties. The concept *molecule* was expressed in an admirable manner by Primo Levi who, in his book *The Monkey's Wrench*, defined the chemist's job by comparing it to an engineer's job. As Levi said:

> *We [chemists] are like blind people with sensitive fingers. I say blind because, actually, the things we handle are too small to be seen, even with the most powerful of microscopes.*

Even if completely blind, chemists have nevertheless been able to demonstrate that there exists a large variety of molecules in Nature, from very simple ones, such as those already

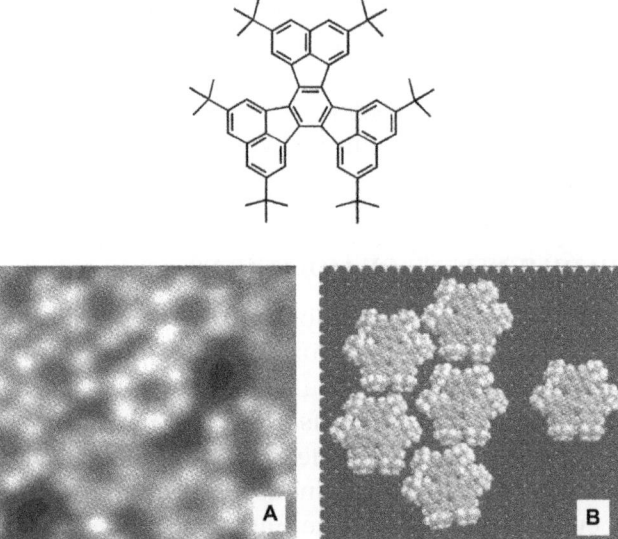

Figure 3.3 Schematic representation (simplified structural formula, top) of the di-esa-tert-butyldecacyclene molecule. (A) Image of molecules of this type obtained with sophisticated microscopic techniques. (B) Computer-generated image that better demonstrates the shape of the molecules.

mentioned—for instance, the oxygen molecule (O_2) and the molecule of water (H_2O)—to the most complex molecules that are found in organisms. Chemists have never given up on their quest *to see* the molecules (see for example Figures 3.3 and 3.4), even though past authoritative people did not share such aspirations.

Goethe, one of the greatest German authors of the 18th century, maintained that science should stick to human dimensions and was opposed to the use of microscopes when he asserted:

that which is invisible to the human eye must not be sought, because evidently it is hidden from the human eye for a very good reason.

This assertion, however, is in sharp contrast to the logic of science, especially in the last few years, which have witnessed investigations of *the small*, not only to better understand Nature, but also to exploit, from a technological point of view, the advantages that could be derived from such knowledge.

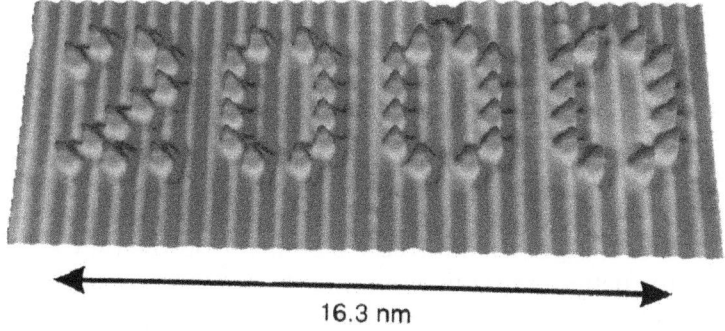

Figure 3.4 Picture of the celebrated date of the new millennium obtained by positioning 47 molecules of carbon monoxide, CO, on a copper surface, taken with a very sophisticated technique known as probe microscopy. Note that the length of the number 2000 is only 16.3 nm long.

3.3 THE NAMES OF MOLECULES

With the hundreds of atomic species available and the various modes with which atoms can bond to each other, it is possible to obtain a large (practically infinite) number of molecules. Therefore, a new world exists that is constituted of extremely small objects (with dimensions of the order of a nanometer), which are incredibly numerous (in a drop of water there are so many molecules that you could distribute some 200 billion to each person on Earth) and highly diversified (the number of molecules known today is in the tens of millions).

Many molecules have common names—for example: water, glucose, hemoglobin, ammonia. However, so as not to get lost in such a diverse and complex world of nomenclature, chemists have found it necessary *to label* the objects in the simplest possible, yet rigorous manner, as is done in classifying plants and animals. Thus, each molecule must be given a non-arbitrary name, a name based on a logical system that expresses, as much as possible, a certain degree of resemblance to other molecules and to its principal properties.

| Names, surnames and addresses |

This is exactly what chemists have achieved by classifying the molecules according to their composition and their properties—for instance, oxides, hydrides, acids, bases, alcohols, ethers, hydrocarbons, proteins, *etc.* They have assigned *surnames* and *names*—for example, methyl alcohol or methanol, CH_4O; ethanol, C_2H_6O; hydrofluoric acid, HF; hydrochloric acid, HCl; hydrobromic acid, HBr—and explained the *relationships* using suffixes and prefixes (perchloric acid, $HClO_4$; chloric acid, $HClO_3$; chlorous acid, $HClO_2$; and hypochlorous acid, HClO). They also specified *addresses* and *civic numbers* (as in the molecule 2-chloro-1-propanol; C_3H_7OCl) so as to better explain how the atoms are connected to each other.

In short, attempts were made to identify and summarize, whenever possible, in one or in a few words, the essence of a specific molecule. However, the large variety of known molecules and the hundreds and thousands of new molecules that are discovered every day in Nature, or otherwise synthesized in the laboratory, render the classification of molecules a daunting task indeed. The extraordinary wealth of the molecular world is difficult to contain in whichever type of linguistic organization one chooses. For the more complex molecules, the tendency now is to abandon the official nomenclature and use, instead, fancy names borrowed from objects encountered in everyday activities. However, we hasten to note that abandoning the official nomenclature will only lead to an undesirable confusion.

3.4 THE FORMULAS OF MOLECULES

Molecular formulas

It is not possible to describe all the properties of a molecule using nomenclature alone. For this reason, the chemist resorts to more adequate methods with which to describe the properties—these are the *molecular formulas*. For instance, H_2O, NH_3 and CH_4 represent the molecular formulas of water, ammonia and methane, respectively. A molecular formula is easy to write and is the simplest document that identifies the molecule, in as much as it tells us which and how many atoms make up the molecule. As an example, the molecular formula of ethanol, a very famous (albeit inflammable) liquid and a substance present in wine and other alcoholic beverages, is written as C_2H_6O; this indicates that the molecule is constituted of 2 carbon atoms, 6 atoms

The World of Molecules

of hydrogen and 1 of oxygen. The molecular formula, however, tells us neither what is bonded to what, nor does it tell us the spatial disposition of the atoms that make up the molecule. Therefore, the molecular formula provides little information regarding the properties of a molecule. Such properties depend not only on the number and the nature of the atoms that make up the molecule, but also on how the atoms are bonded and how they are distributed in three-dimensional space.

Structural formulas

The formulas that illustrate the bonds that connect the various atoms in a molecule are known as *structural formulas*; some examples are given in Figure 3.2. Although such structural formulas depict what is bonded to what, they still fail to explain the distribution of the atoms in space. Such spatial distribution of atoms comes into play in triatomic molecules since the three atoms can either all lie on the same line, or can form an angular structure. If the structure of the molecule of water, H_2O, were linear, for example, water would not have the properties we know it has, as such properties descend from its angular structure (Figure 3.5). Scientists have demonstrated that if the water molecule possessed a linear structure, there would be no ice and no water on Earth, only water vapor, in which case life could not have evolved as we know it!

When there are more than three atoms in a molecule, their disposition in space can vary in three dimensions: the atoms can be on the same line, they can lie on the same plane, or else they

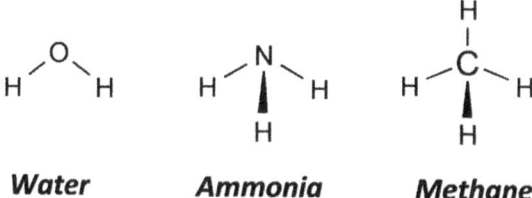

Water   **Ammonia**   **Methane**

Figure 3.5   The atoms that constitute a molecule are disposed at fixed positions in three-dimensional space. The water molecule has an angular structure; the structure of ammonia is pyramidal, whereas methane has a tetrahedral structure.

can be arranged in a three-dimensional structure. Chemists have shown, for example, that the molecules of ammonia, NH_3, and of methane, CH_4, are not planar but form, respectively, a pyramid and a tetrahedron. Such structures are typically represented using the structures of geometric solids.

> **Spatial disposition of the atoms**

Let's consider now the example of ethanol, whose molecular formula is C_2H_6O, in more detail. This is a molecule that contains a more complex bond network; its structural formula is displayed in Figure 3.6. Chemists know that the essential properties of ethanol, a liquid at ambient temperature, are simply due to the type of bonds between the atoms of the molecule and, particularly, on the fact that it contains an O–H group; that is, an atom of oxygen is bonded to a hydrogen atom.

> **Isomerism**

In the course of their studies, chemists also noted that there is another substance that has the same molecular formula as ethanol, C_2H_6O, but with different properties. This is the molecule known as dimethyl ether—a gaseous substance with a strong odor—which possesses anesthetic properties (Figure 3.6). Molecules with the same molecular formula but with different properties are referred to as *isomers*. In the case just mentioned, the different properties are attributed to the different ways in which the atoms are bonded in the respective molecules, that is, to different structural formulas (isomeric structures).

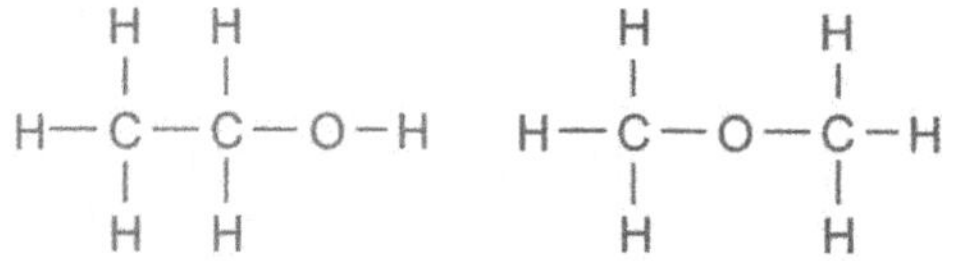

```
      H   H                    H       H
      |   |                    |       |
  H — C — C — O — H        H — C — O — C — H
      |   |                    |       |
      H   H                    H       H

     Ethyl alcohol             Dimethyl ether
        C₂H₆O                     C₂H₆O
```

Ethyl alcohol
C_2H_6O

Dimethyl ether
C_2H_6O

Figure 3.6 The molecules of ethyl alcohol (ethanol) and dimethyl ether possess the same molecular formula, C_2H_6O, but different structural formulas.

The World of Molecules

Analogously, this also happens in language. Just like molecules, the meaning of a word depends not only on which and how many letters form the word, but also on the order in which the letters are written. As an example, the two Italian words *giravolta* and *travaglio* are formed from the same letters. Yet their meaning is very different (*twirl* versus *anguish*) in virtue of the fact that the constituent letters are ordered differently.

Naturally, the problem of isomerism becomes even more complex as the number of atoms in molecules increases, and also because molecules, which have the same chemical composition and atoms bonded in the same manner, are different if they have different structures. Such an occurrence is referred to as *stereoisomerism*.

> **The same and not the same**

Figure 3.7 exhibits the case of alanine, a member of the family of amino acids. In this molecule, the central carbon atom is bonded to a nitrogen atom, a hydrogen atom, and two other carbon atoms. However, the latter two carbon atoms are not equivalent to the central carbon atom because they are bonded to different atoms. In fact, alanine is one of the many substances that exist in two forms in the biological world. Although the physical structures of these two forms are identical (tetrahedral about the central carbon atom; Figure 3.7), they are spatially different. Such substances are referred to as *chiral isomers*, sometimes also called *optical isomers*. This can easily be shown if we look at our hands: the right hand is the mirror image of the left hand. That they are different spatially is demonstrated by the fact that we cannot put the glove of the left hand into the right hand, and vice versa.

Spatially different structures of this type that we could classify as useless details are, in fact, of fundamental importance in the world of biological molecules. For example, the molecules of many drugs exhibit the same structural problems as alanine. It may happen that one of the two molecular structures may turn out to be beneficial, while the other may turn out to be quite toxic, indeed lethal.

On this subject, some of us folks will no doubt remember the case of the drug known as *thalidomide*. It was used against nausea and to alleviate morning sickness in pregnant women. In the late 1950s and early 1960s, when taken by pregnant women,

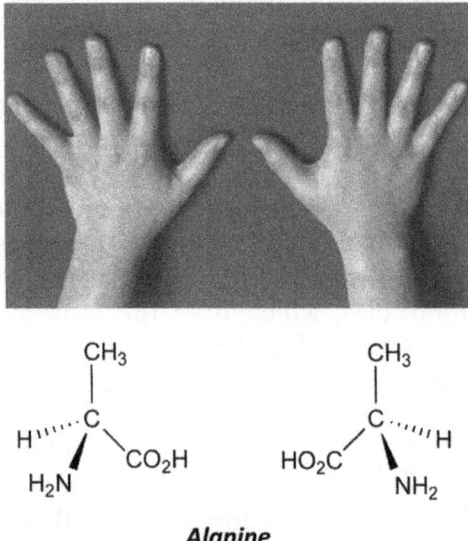

Figure 3.7 Alanine is an example of a molecule that can exist in two forms. Such apparently identical (the same) molecules are in reality spatially different (not the same), like our two hands. The two forms of molecules of this type (chiral isomers or optical isomers) tend to display drastically different properties. In the case of alanine, only one of the two forms is present in organisms.

the drug was shown to be the cause of birth defects in many children, who were born with badly formed limbs or otherwise with no limbs at all. The drug was commercialized during that period as a mixture of its two chiral isomers. It was only a few years later that scientists discovered that when taken by a pregnant woman, one of the two chiral forms of thalidomide had devastating effects on the development of the embryo. For this reason, thalidomide was withdrawn from the market in 1961 in the United Kingdom and later in other countries.

3.5 MOLECULAR MODELS

Scientists have found molecular formulas and structural formulas very useful. However, it cannot be said that such formulas are eye-catching. In fact, too often these formulas engender great difficulties for students, particularly when they are introduced in a boring and mechanical fashion, and without any connection to the real world of molecules they are meant to represent. Only people that possess some knowledge of chemistry will likely appreciate the structural formula of a newly discovered or newly synthesized molecule. Observing the structural formula carefully, in fact, the expert viewer is in a position to read many properties of the substance that the formula represents: for example, whether it is soluble in water, whether it is an acid or a base, whether it is potentially explosive, or whether it is colored, *etc.*

For small or relatively small molecules, the structural formulas are rather simple and, in addition to indicating clearly how the atoms are bonded, they also provide very useful information to the chemist. However, as we move to larger and larger molecules the situation becomes more complex and the structural formulas begin to resemble an intricate network of symbols. Accordingly, attempts were then made to simplify such formulas; for example, the carbon atoms, C, which are commonly found in molecules, especially those of living organisms, are no longer indicated explicitly by the symbol C, but are understood to occupy the positions at the points where the (horizontal) dashes—that is, the bonds—intersect. Also, the hydrogen atoms, H, bonded to the carbon atoms are no longer displayed and their bonding to the C atoms are also no longer shown explicitly (see Figure 3.8 for some examples).

Notwithstanding such simplifications, the structural formula of a complex molecule gives no information as to the dimensions of the molecule, its shape and, most of all, how the different atoms of the molecule are spatially disposed.

The Lego game of molecules

The double inabilities of formulas to represent dimensions and shapes of molecules are compensated by the use of three-dimensional models. These models are constructed by means of a method that resembles the famous game *Lego*, beginning with small balls of rigid plastic that represent the several types of atoms supplied with small cavities into which sticks can be inserted. The sticks represent the chemical bonds. Every ball that represents an atom is a hundred million times larger than the real atom, so that the model is to scale and thus faithfully represents the relative dimensions of various molecules, together with the parts that make

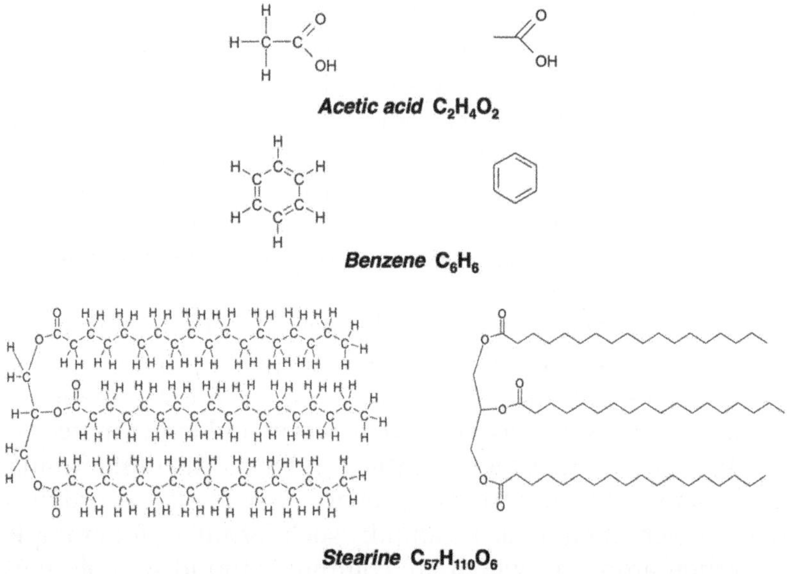

Acetic acid $C_2H_4O_2$

Benzene C_6H_6

Stearine $C_{57}H_{110}O_6$

Figure 3.8 Instead of illustrating the entire structural formulas (left hand side), chemists often use simplified representations (right hand side), particularly in the case of complex molecules.

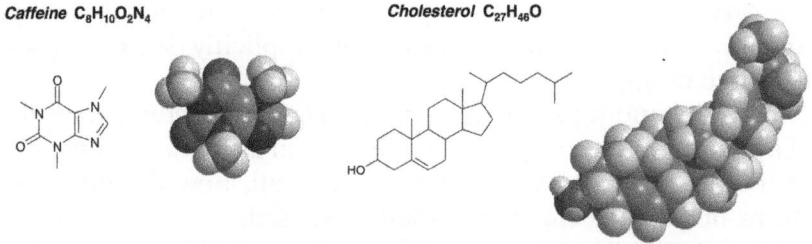

Caffeine $C_8H_{10}O_2N_4$

Cholesterol $C_{27}H_{46}O$

Figure 3.9 Molecular formulas, simplified structural formulas and three-dimensional models of the molecules of caffeine and cholesterol. Note that the proportions between the two molecules have not been respected in the image.

up the molecules. In order to distinguish the several types of atoms or, better still, the most recurrent ones in important molecules, conventional colors are used: white for hydrogen (H), black for carbon (C), red for oxygen (O), blue for nitrogen (N), yellow for sulfur (S), and green for chlorine (Cl). Of course the different atoms are not so clearly shown in our black and white images. Nevertheless, Figures 3.9–3.11 show that using three dimensional models the molecules regain part of the fascination that they would have had if we could see them and touch them in the real world.

3.6 ATTENTION TO DETAILS!

Even though it is an object with nanometer dimensions, every molecule possesses a characteristic composition, dimension, structure and shape. Moreover, in every molecule the various atoms are bonded to each other so as to constitute a network with very particular relationships. From all this, we can deduce that every type of molecule has well-defined specific properties, and that molecules of the same type have the same properties. Nevertheless, a small difference in composition, dimension, shape and distribution of the atoms or of the bonds suffices to radically modify the properties of a molecule. Accordingly, let's consider some examples.

Red and blue

The red color of the poppy and the blue color of cornflower are caused by the same molecule that belongs to the family of anthocyanins. This diversity in color results from the fact that the molecule can lose a hydrogen ion, H^+. In acidic media, the molecule contains the H^+ ion and is red in color, while in alkaline media the molecule no longer contains the H^+ ion and consequently displays a blue color. The poppies that have an acidic sap are red; the cornflowers that have an alkaline sap are blue. The experimental proof to test this assertion is to make an infusion of the petals of poppies with ethyl alcohol (ethanol). If the resulting red solution is treated with a base (an alkali) it turns blue; that is, it assumes the color of the cornflower.

Male and female

The molecule of the male sexual hormone, testosterone, and the molecule of the female sexual hormone, estradiol, have nearly identical structures (Figure 3.10). They differ ever so slightly such that only a careful observer is able to note the difference (this could be a case for an enigmatic game).

The eye and the light

The way in which our eyes see things is due to the light-induced transformation of the geometric

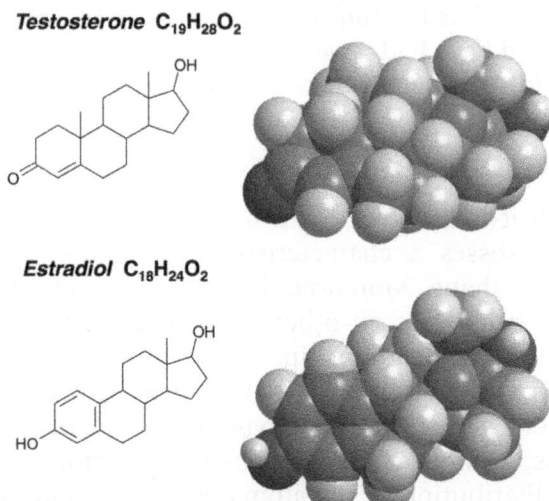

Figure 3.10 Molecular formulas, simplified structural formulas, and molecular models of the male/female sexual hormones: testosterone (male) and estradiol (female).

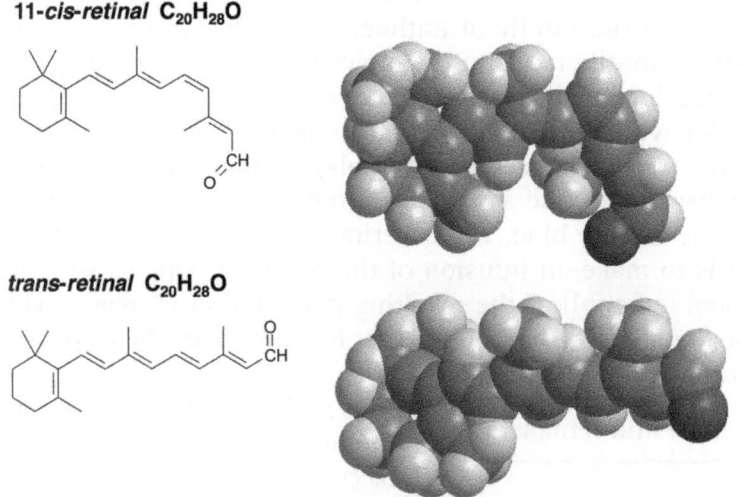

Figure 3.11 The two isomers *cis* and *trans* of the retinal molecule that plays a leading role in the vision phenomenon.

structure of the retinal molecule (Figure 3.11). Light changes the geometric form of this molecule from its *cis* isomeric form to its longer form, the *trans* isomer.

CHAPTER 4
Chemistry in Action: The Reactions

4.1 TRANSFORMATION OF CHEMICAL SPECIES

The transformation of one or more chemical species into another species is said to be a chemical reaction. In this transformation, the atoms in the initial species—the *reagents*—undergo a rearrangement amongst themselves; that is, they undergo bond breaking and bond making so as to form new species: the *products*.

Besides being processes that can be performed in a chemical laboratory, chemical reactions are, above all, events that take place continuously everywhere we look. For instance, items made of iron tend to rust, bread dough rises because of yeast fermentation, and the wine turns into vinegar by air oxidation. These are all examples of chemical reactions occurring in everyday life. Even the conception of life (the actual birth) and the termination of life (death) are all the result of a complex sequence of chemical reactions.

| **Nothing is created or destroyed, but all is transformed** |

The atoms and molecules we are made of are in continuous exchange with those of our environment through chemical reactions. When we breathe, eat and drink, we take

---

Chemistry: Reading and Writing the Book of Nature
By Vincenzo Balzani and Margherita Venturi
Translation by Nick Serpone
© The Royal Society of Chemistry 2014
Published by the Royal Society of Chemistry, www.rsc.org

on billions and billions of atoms from the environment. Then, we reject as many back into the environment through sweating, exhaling, and eliminating biological wastes. Though it may seem a little poetic, with every breath we capture, we then put back into the environment some billions and billions of atoms that had already been recycled in the last few weeks from the breaths of other living beings. Everything in us is continuously renewed as we draw from matter and energy in our environment. Our skin renews itself every month, our liver every six weeks. Every year, 98% of our body is renewed. Consequently, we can safely say that, in the real world, we are the most recycled species. Even our memories, which are particular structures of the brain, are continuously being dismantled and reassembled through atomic and molecular exchanges.

A chemical reaction can occur either *silently* with no evidence of change, or else it may be accompanied by spectacular phenomena, such as substantial changes in color or an explosive evolution of a gas. Very common are combustion reactions in which a substance—the combustible—reacts with oxygen—the combustive agent. Without searching for bizarre examples, suffice it to note the burning of a candle. In this regard, Michael Faraday, a chemist and physicist of 19th century England, liked to begin his famous chemistry lectures by analyzing the flame of a lit candle (Figure 4.1).

Reactions around us and in us

Even precipitation reactions are widespread. Other than the classical laboratory experiments, such as the precipitation of silver chloride (AgCl; white color) we only need to look around us to discover vast choices: the precipitation of salt that takes place in saline marshes by evaporation of sea water, formation of clam shells, and formation of beautiful stalactites and stalagmites in certain caves.

Barium sulfate ($BaSO_4$) is a toxic substance that absorbs X-rays and is water-insoluble. Being water-insoluble makes $BaSO_4$ suitable as a contrasting agent when taking X-ray images of the digestive apparatus, inasmuch as its complete insolubility prevents the substance from displaying the relatively inherent high toxicity of the barium ion. Here is an example, then, that demonstrates the relevance of the solubility of salts on possible biological effects.

Figure 4.1 Michael Faraday (1791–1867) explaining the combustion of a candle to a public audience at the Royal Institution in London, England.
Reproduced from http://lowres-picturecabinet.com.s3-eu-west-1.amazonaws.com/43/main/8/87643.jpg.

Such precipitates as silver chloride (AgCl) can be dissolved through suitable reactions. Indeed, the solubilization of AgCl can be achieved by the addition of ammonia (NH_3), which results in the formation of a water-soluble species (referred to as a complex) between the silver ion and ammonia, namely $[Ag(NH_3)_2]^+$. The solubilization of calcium carbonate ($CaCO_3$) occurs by addition of an acid, resulting in the formation of carbon dioxide (CO_2) which, being a gas, is released into the environment. The latter reaction is particularly interesting because it demonstrates the effect that acid rain has on monuments made of marble.

Equally interesting are the isomerization reactions, not only for their biological implications, but also for their industrial applications: for example, the photochromic materials that change color upon being exposed to light.

Despite the fact that all chemical reactions must be considered important, undoubtedly the oxidation–reduction reactions (or so-called *redox reactions*) and those that involve acids and bases are of particular interest from the biological point of view, and for their usefulness in applied fields.

4.2 OXIDATION–REDUCTION (REDOX) REACTIONS

Oxidation–reduction reactions are of great importance in both the natural world and in the various fields of human activity. A few examples are worth noting:

- the loss of the sheen of silver;
- the browning of a slice of apple exposed to air;
- photosynthesis, which consists of a most complicated sequence of oxidation–reduction processes that are triggered by light.

Electron transfer

Redox reactions imply the removal of electrons from one substance and their capture by another. The species that loses the electrons is said to be oxidized, while the species that acquires the electrons is said to be reduced. An oxidation process is always associated with a process of reduction, as electrons are neither created nor destroyed, but are simply transferred from one species to another. This transfer of electrons can be demonstrated experimentally by connecting—through an external circuit—the species that is oxidized to the one that is reduced in separate vessels but communicating with each other by means of a salt bridge. The spontaneous electron transfer from the species oxidized to the species reduced is observed experimentally by a flow of current in the external circuit.

The batteries

Electron transfer is precisely the principle on which batteries, so essential in modern society, operate. They supply energy, the greatest consumer product, in more and more compact forms. For instance, lithium batteries in cellular telephony have rendered possible the miniaturization of many ultra-light cell phones currently used, and have made it possible to *maintain the battery charge* for several hours.

The Italian scientist Alessandro Volta (1745–1827) was the first to fabricate the battery (or *pila*) (Figure 4.2). The word *pila* originates from the fact that it was obtained by stacking up (or piling up) disks of copper and zinc, interspaced with small disks of carton impregnated with an acid.

Combustion reactions

Combustion processes are important oxidation–reduction reactions

Figure 4.2 Alessandro Volta and his "pila" (battery).

that convert the chemical energy stored in fuels (coal, gasoline, methane, but also wood, *etc.*) into thermal energy (heat). In a sense, this represents a chemical process in complete opposite to the process of photosynthesis, inasmuch as by consuming oxygen it converts compounds rich in energy (for example, the hydrocarbons in gasoline) into compounds low in energy content (water and carbon dioxide). The thermal energy so obtained can be used either as such (heating houses, cooking of foods) or can be converted subsequently into other forms (mechanical energy or electrical energy). Though being a spontaneous process, combustion does not occur unless it is triggered by an external stimulus (a lit match to the gas stove, a spark in the internal combustion engine, *etc.*). This distinguishing feature is of particular interest because it allows preserving substances that have an elevated energetic content until the moment of their use.

Respiration is, so to speak, the biological version of the combustion process and, like all the biological processes, is much more complex. In essence, in the respiration process the substances with a high energetic content within foodstuffs react with oxygen and are subsequently transformed into products of lower energy content (carbon dioxide, water and others), thereby supplying the living organism with the necessary energy to support life.

4.3 ACID–BASE REACTIONS

The notion of acids and bases has changed considerably over time with the development of various theories. Without going into detail, acids and bases can be defined in a simple, yet substantially correct fashion. Acids are molecules or ions that, in water, release hydrogen ions (H^+, also referred to as a proton, which associates with a water molecule giving rise to the hydronium ion, H_3O^+). By contrast, bases are molecules or ions that extract a proton from a water molecule giving rise to OH^- ions. Examples of acids are: hydrochloric acid (HCl, also known as muriatic acid and present in gastric juices); sulfuric acid (H_2SO_4, once referred to as vitriol), and several foodstuffs such as yogurt, vinegar and lemon. Examples of bases (alkali) are such substances as ammonia (NH_3, used domestically as a degreasing agent), and potassium hydroxide (KOH) present in liquids used to unclog blocked kitchen and bathroom sinks.

Proton transfer

Of great interest are the neutralization reactions that occur when an acid reacts with a base. Such reactions reveal the antagonistic effect of these two types of substances. For this reason, sodium bicarbonate ($NaHCO_3$), a substance with basic (alkaline) properties, is commonly used to counteract the excess acidity in the stomach.

Acids and bases have significant implications in the environmental field (acid rain), in the industrial field (for example, the acidity of tomato sauce must be controlled closely so that its pleasant flavor is maintained) and, above all, in the biological field. Our organism, in fact, acts like a perfect and efficient chemical factory that automatically controls and constantly adjusts the acidity of the blood and other physiological liquids. To the extent that the level of acidity governs nearly all other reactions that take place in our body, were this control to lessen, death would occur within a short time.

Acids and bases are also very useful in controlling and steering many chemical reactions. The products obtained from two very specific reagents can, in fact, change drastically by varying the acidity of the reaction environment. A particularly interesting example of this behavior is the reduction of the permanganate ion (MnO_4^-) in aqueous media, as it involves an observable change of color. If this intensely violet-colored ion were to be reduced in acidic media, it would lead to the formation of the pale pink manganese ion (Mn^{2+}). However, if the reduction were carried out in neutral or in slightly alkaline media, the product formed would be manganese dioxide (MnO_2), which appears in the form of a black precipitate. Finally, if the reaction occurred in a strongly alkaline environment, it would yield the emerald green manganate ion (MnO_4^{2-}).

Quantifying acidity and alkalinity: pH

When describing acids and bases, we encounter the concept of pH. The term pH has now become a part of the common lexicon. Phrases such as *it does not alter the pH of the skin* or *it has a neutral pH* can often be read in the printed media or otherwise heard spoken. Leaving aside its exact definition, pH can simply be said to be a measure of the concentration of the hydronium ion (H_3O^+). The parameter pH was chosen because it also gives an indirect measure of the concentration of the hydroxide ion OH^-. This is possible since the concentrations of these two ions in aqueous solution are tied to the fact that their product must be constant: $[H_3O^+] \times [OH^-] = 10^{-14}$. That is, if the amount of one decreases, then the concentration of the other must necessarily increase.

For operational reasons, it is appropriate to remember that:

- when the pH is smaller than 7, the concentration of the H_3O^+ ion prevails and the solution is said to be acidic;
- when the pH is greater than 7, the concentration of the OH^- ion prevails and the solution is said to be basic or alkaline;
- when the pH is exactly equal to 7, the concentration of the H_3O^+ ion is equal to the concentration of the OH^- ion and the solution is said to be neutral (pure water is neutral).

Color as an indicator of pH

An important aspect of acids and bases concerns the possibility of determining experimentally the pH and therefore the concentration of the hydronium ion H_3O^+. This can be obtained quantitatively, with great precision, using a suitable instrument (the pH meter), or more qualitatively, by means of appropriate substances known as *indicators*. These indicators assume a different color according to the level of acidity of the solution. From the variation in color of an indicator, it is possible to show either the presence of an acid or the presence of a base in water, and to demonstrate that a base neutralizes the effect of an acid (and vice versa). It can also be shown that some salts do not alter the pH of water, for example, sodium chloride (NaCl, a common salt found in many kitchens). Other salts behave as acids—for example, ammonium chloride (NH_4Cl), which is often used as yeast—while others act as bases, such as sodium hypochlorite (NaClO, commonly known as bleach or as javel water in North America) and sodium bicarbonate ($NaHCO_3$).

Many compounds in Nature change color when the acidity varies. Common examples are tea and red wine, as evidenced by the change of color on the addition of lemon juice. Other examples are substances that impart color to flowers and fruits. For instance, the juice of red cabbage displays an extraordinary spectrum (range) of colors, with changes in the pH of the solution.

4.4 CHEMICAL EQUATIONS

The symbolic level of chemistry

Chemical equations are the symbolic and synthetic representations of chemical reactions. Through the chemical formulas of reagents and products, they give qualitative information on the type of reaction, and through the numbers placed in front of every formula they also give quantitative information. These numbers are the stoichiometric coefficients that serve to balance the chemical equation. As an example, for the equation reported in Figure 4.3 the stoichiometric coefficients are: 1 for the reagent CH_4 and product CO_2 (for simplicity this coefficient is implied) and 2 for the reagent O_2 and product H_2O.

Material balance of reactions

In balancing a chemical reaction, it must be emphasized that chemistry is also the science of Nature's book-keeping. Balancing chemical reactions is, in fact, a process of material balance: all the atoms on the left hand side of the equation must be found again on the

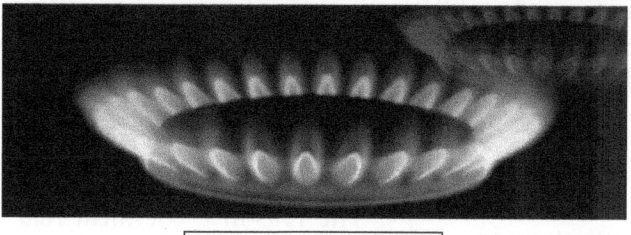

$$CH_4 + 2O_2 \rightarrow CO_2 + 2H_2O$$

Figure 4.3 Example of a quantitative balanced reaction: the combustion of methane.

$N_2 + O_2 \leftrightarrows 2NO$ $H_2CO_3 \leftrightarrows CO_2 + H_2O$

Figure 4.4 Two examples of balanced incomplete reactions: the transformation of nitrogen and oxygen into nitrogen oxides that occurs in the atmosphere during thunderstorms, and the development of gas bubbles (CO_2) when gaseous drinks are uncorked.

right hand side, because matter cannot disappear in a chemical reaction!

Chemical reactions can be either quantitative or incomplete, which, in the symbolic representation of the process, is indicated by placing either an arrow (Figure 4.3) or two arrows in opposite directions (Figure 4.4) between the formulas of the reagents and those of the products. The difference between these two types of reactions consists in the fact that, in the first case, the transformation of the reagents into products is complete (quantitative reactions), while in the second the transformation is only partial (incomplete reactions).

The quantitative information gained from a balanced chemical equation can be used to predict the amounts of products formed and the quantities of reagents transformed in a specific process. Such predictions have important implications in applied fields.

4.5 WHY REACTIONS OCCUR

Spontaneity in chemistry

The reason a chemical reaction occurs finds its justification in the principles of thermodynamics. The *spontaneity* of a chemical reaction, however, that is the tendency of reagents to transform themselves into products, can also be described in a simple manner. At first, the problem can be set up considering some simple physical phenomena with which we are continuously in contact, and for which spontaneity is a rather obvious concept. For instance, if a ball were placed on top of a hill, the ball would roll spontaneously toward the bottom of the hill. In the field of electricity, a positive charge is spontaneously attracted to a negative charge. We would be surprised if we saw the ball "roll up" the hill, or the positive charge repelled by the negative charge. Spontaneous processes are those that tend toward a decrease of some form of energy of the system. In the case of the ball, the energy is of the gravitational type, while in the case of the electrical charge, it is of the electrostatic type.

Energy and disorder

Considering processes more closely related to chemistry, such as phase transformations, that is, passage between the states of aggregation (solid, liquid, gas), we note that in many cases they take place by the release of heat (decreasing the energy of the system). However, it cannot be said that only processes that release heat are spontaneous. In fact, common experience tells us that many spontaneous transformations occur by absorption of heat. As an example, evaporation of a liquid requires heat in order to overcome the forces of attraction between the molecules of the liquid. Recall that water in an open container evaporates spontaneously with the passage of time.

Similarly, many chemical reactions occur spontaneously, even though they require heat absorption, that is, even in a situation involving an increase in the energy of the system. Clearly, energy cannot be the sole factor that determines the spontaneity of a process. A closer analysis of the spontaneous processes that require heat reveals that they proceed toward a situation of increased disorder. As far as evaporation of water is concerned, the process involves passage from the liquid state, in which molecules are forced to be near each other because of intermolecular forces, to the gaseous state, in which the molecules are free to move about in a completely disordered fashion. Even for chemical reactions, we observe that, among processes that take place with heat absorption, only those that lead to a more disordered condition are spontaneous. For example, this is the case of a process that, beginning with reagents in the solid state, in which the crystalline network imposes an extreme order, leads to the formation of one or more products in the liquid or gaseous state characterized by a greater molecular disorder.

Thus, the spontaneity of a chemical reaction depends on two factors:

- release of heat;
- increase of disorder.

The exchange of heat (thermal energy) between a system and its environment, and the variation of the degree of disorder in the system sometimes cooperate. More often, however, they conflict with each other, so that the spontaneity of the reaction depends only on the balance between energy and disorder. Reactions that take place with release of heat and with an increase of disorder are always spontaneous. Reactions that occur with absorption of heat, but with an increase of disorder, are spontaneous only if the positive effect of increasing disorder exceeds the negative effect of heat absorption. Similarly, reactions that occur with release of heat, but with an increase of order (*i.e.* decrease of disorder), are spontaneous only if the positive effect of the heat release exceeds the negative effect of order increase. Finally, reactions that involve absorption of heat and an increase of order never occur spontaneously. In the latter case, we emphasize that the assertion *never occur spontaneously* does not infer that they *never* happen in the absolute sense. In fact, it is possible to find a way for a reaction to proceed spontaneously through the participation of external factors, even though the reaction itself may not be spontaneous.

It is appropriate to ask why is it that in the last three billion years there has been continuous progress from the simplest compounds to more complex and organized systems until the arrival of humans, if everything tended spontaneously toward a situation of lesser energy and maximum disorder. Consequently, it seems rather strange that a complex phenomenon such as life, so perfectly organized, could have developed on Earth. This apparent contradiction crumbles from the moment we recognize that, in fact, reactions that sustain life do not occur in isolation (that is, all alone), but draw energy continuously from the outside world so as to sustain itself. Our source of energy is the Sun. As long as the Sun shines in the sky, which will occur for a long time to come, life will continue to exist on Earth.

4.6 REACTIONS AND TIME

> **Spontaneity is not a guarantee that a reaction occurs**

The spontaneity concept does not consider the time factor. Experience has shown, however, that the time taken for a spontaneous reaction to occur is one of its intrinsic characteristics, which varies from case to case. Undeniably, there are fast spontaneous reactions (Figure 4.5) that take place within a few milliseconds (ms), and then there are spontaneous reactions that might take 100 years to occur to any appreciable extent.

It is possible to influence the rate (speed) of a spontaneous reaction operating under certain conditions. Accordingly, it is important to understand, at least qualitatively, *how* one can influence the rate of a reaction.

> **How to influence the rate of a reaction: the temperature factor**

Temperature is clearly the first factor to take into consideration. It is easy to demonstrate the influence of temperature on the time needed for a reaction to occur, considering that this effect is continuously exploited, often involuntarily, in daily activities. For example, foods are cooked so as to accelerate the desired reactions which, by causing the breakup of cellular walls and the decomposition of some proteins, confer aromas

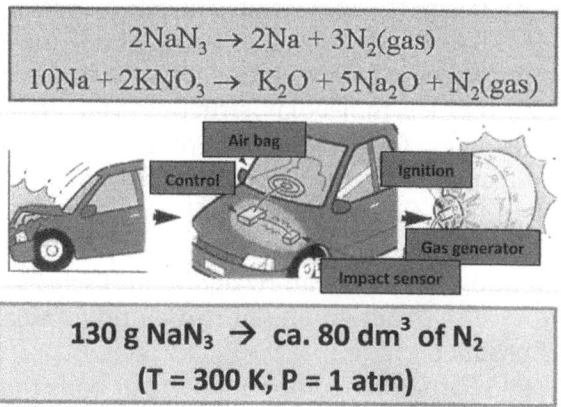

Figure 4.5 The more common types of air bags function thanks to a reaction that produces very rapidly, in fact explosively, a large quantity of the innocuous gas nitrogen. The nitrogen so produced inflates the air bag in a very short time, thereby protecting whoever is in the car from the consequences of a violent collision.

and pleasant flavors to foods, while helping the digestion. In contrast, foods are kept in a refrigerator precisely because undesired reactions (fatty foods turn rancid; proliferation of mildew) are slowed down at lower temperatures. Acceleration of a process by simple manipulation of the temperature is of great importance in industry.

How to influence the rate of a reaction: the use of catalysts and inhibitors

The other method to accelerate a reaction involves the use of *catalysts*. Although this method is less intuitive than the temperature factor, it is nonetheless equally important and one that is exploited extensively, not only industrially, but also by Nature. Catalysts are substances that participate in the reaction but are not consumed because their task is to facilitate the path the reagents must take to be transformed into products. As you would expect, it is impossible to find a *universal catalyst*, that is, to find a substance that is in a position to accelerate different types of reactions. The contrary is in fact true. The more the catalyst is specific for a given type of reaction, the greater is its effectiveness in accelerating that reaction.

Our body is an expert in the choice of catalysts it needs. Every process is guided by appropriate substances—the *enzymes*—which carry out specific catalytic functions. Enzymes are species possessing a complex molecular structure and display an unbelievable efficiency and selectivity. For example, they can increase the rate of a process by several orders of magnitude (up to a thousand billion times faster), and can operate effectively under mild conditions of temperature and pH. In the case of chiral compounds, enzymes can recognize and selectively choose the reagent, or else generate one of two chiral (optical) isomers as the final product.

However, there are other compounds in our body that function either to prevent or to slow down some undesired processes. These are known as *inhibitors*. To understand their importance we need only remember the anti-oxidants such as, for example, vitamin C which, among other things, slows down the oxidation reactions responsible for the aging process.

Choosing effective catalysts and specific inhibitors is of extreme importance in the industrial field. Inhibitors can avoid or slow down the degradation of materials. Catalysts can reduce the cost of synthetic processes by reducing reaction times, increasing production yields, and allowing operating at lower temperatures. In the pharmaceutical industry, for example, the availability of catalysts that can impact a synthesis toward the production of a single chiral isomer is important as many of the drugs are of the chiral type and often one of the two isomers is pharmacologically inactive, if not quite toxic. Among the many applications of catalysts, we need only recall the catalytic converters in automobiles. These catalysts contain small amounts of noble metals (platinum and rhodium) that convert the nitrogen oxides (so-called NOx compounds), carbon monoxide and the hydrocarbons that escaped combustion into less harmful exhaust substances.

CHAPTER 5

Beyond Molecules: From Chemistry to Biology

5.1 FROM MOLECULES TO SUPRAMOLECULAR SYSTEMS

Molecules represent the first fundamental step in the scale of chemical complexity (Figure 5.1). In fact, molecules constitute more complex systems than atoms. The term "complex" is not synonymous with complicated. Rather, it refers to systems consisting of more than one entity; in this case, atoms interacting with each other create novel properties.

In the specific case of molecules, the new properties consist of a system with well-defined composition, of a particular shape, and of a given topology, among others. The ensemble of these properties constitutes the *information content* of every molecule. This is important not only because it identifies the molecule, but also and most importantly because it determines the consequence of its interactions with other molecules.

| Association of molecules |

When two molecules encounter each other, each molecule reads the information elements contained in the other molecules and, depending on such components, they either ignore each other, react to produce new species, or else associate to form a supramolecular

---

Chemistry: Reading and Writing the Book of Nature
By Vincenzo Balzani and Margherita Venturi
Translation by Nick Serpone
© The Royal Society of Chemistry 2014
Published by the Royal Society of Chemistry, www.rsc.org

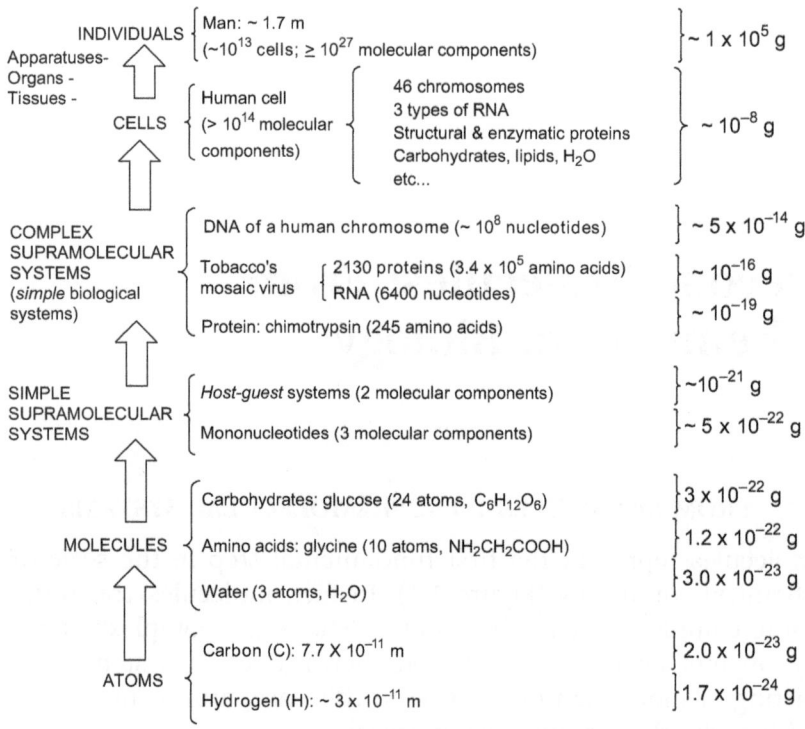

Figure 5.1 Schematic representation of the scale of chemical complexity. Some examples are reported for each level with their salient characteristics. For the simple supramolecular systems, reference is made to the *host-guest* complex illustrated in Figure 5.2b and to the mononucleotides of DNA displayed in Figure 5.3. For the complex supramolecular systems, we have chosen as examples, the simplest enzymatic protein (chymotrypsin), one of the viruses examined in greatest detail (the mosaic virus of tobacco) and the most important biological polymer (DNA). Insofar as the human cell is concerned, only the fundamental components are indicated. The figure highlights the unbelievable complexity of the "chemical system", man, even when analyzed only from a quantitative point of view (one billion billion billion molecular components).

system. Association between molecules occurs through a phenomenon known as *molecular recognition*, a highly specific and selective interaction comparable to lock-and-key recognition. Figure 5.2 illustrates two such examples of molecular association based on the formation of hydrogen bonding; this type of molecular recognition is particularly significant to the biological world. Note that the hydrogen bond is considerably different from the typical covalent bond described earlier.

The first example in Figure 5.2a demonstrates how a molecule of cytosin associates itself with a molecule of guanine, thanks to the formation of three hydrogen bonds. The second example (Figure 5.2b) shows how the small barbital molecule gets, so to speak, enclosed in another molecule that possesses a made-to-measure cavity that can accommodate it (a *host-guest* system).

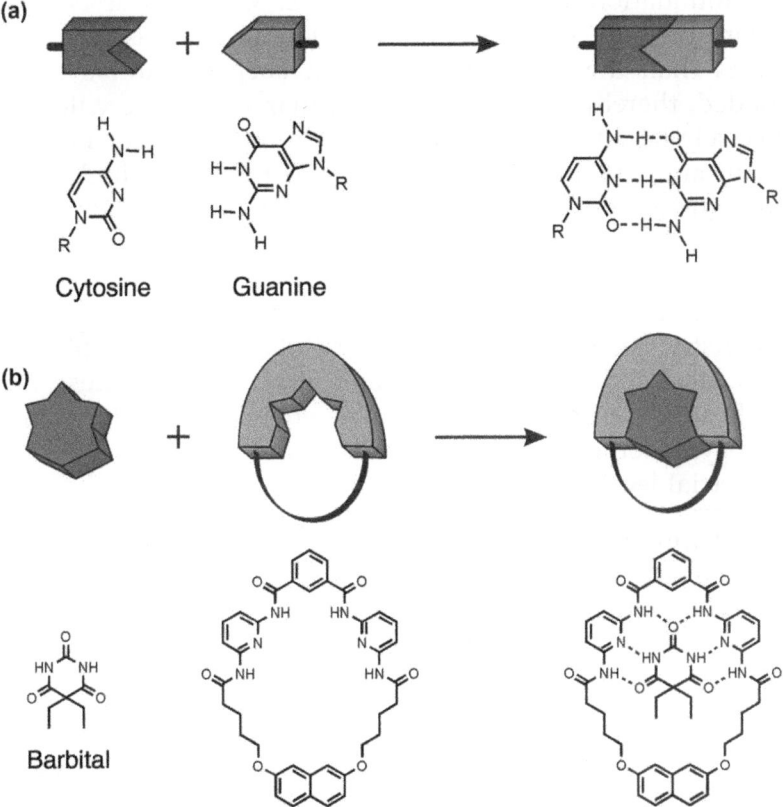

Figure 5.2 Examples of molecular recognition based on the formation of hydrogen bonds: (a) association between cytosine and guanine, two principal nitrogen bases present in DNA; (b) association between barbital and a macrocyclic molecule of suitable dimensions. Also given for each example is a schematic representation that recalls the specificity of an interaction, the lock-key model.

The cartoons of each example in Figure 5.2 underline the analogy with the lock-and-key system. To associate themselves, the molecules must have a perfect structural complementarity. This means that molecular recognition can be programmed through *codes* based on the specific localization of certain atoms (or group of atoms), on the shape, and on other electronic/structural features of the molecule.

A large number of molecules are present in Nature. They are so programmed as to undergo association from which originate supramolecular systems involved in processes that lie at the very foundation of the evolution of life. Molecular association occurs through much weaker interactions between the molecules than the covalent bonds that keep the atoms tightly bonded, thereby making supramolecular systems very flexible. Precisely because they are weak, these interactions can easily occur and are broken up, thereby correcting initial errors through successive attempts, and exploring new structures that could prove very useful from an evolutionary point of view. We must not forget that progress in the various fields of chemistry, and in particular the birth of supramolecular chemistry, has led to the synthesis of a large variety of planned molecules and artificial supramolecular systems. These find application in several segments of chemistry, medicine and biology, whether at the fundamental research level or at the industrial level.

> **Nature makes extensive use of molecular association**

Supramolecular systems represent a step up in the scale of chemical complexity; that is, such systems are one level up from molecules. Even in this case, what matters is not so much an increase in structural complexity. Rather, it is mostly in the notion that, when going from simple molecules to supramolecular systems, properties emerge that are not otherwise present, even conceptually, in separated molecules. In fact, because of the interactions established between molecules, supramolecular systems possess superior information contents relative to those of molecules that make up the supramolecular system. Accordingly, supramolecular systems can perform functions that are not otherwise possible by the component molecules.

5.2 FROM SUPRAMOLECULAR SYSTEMS TO CELLS

The properties that emerge in simple supramolecular systems, formed from a small number of components, are easily predictable from those of the separated molecular components. However, as the number of components that form the system increases, the new characteristics turn out to be more and more difficult to interpret. We go from an intermediate situation in which the emergent properties are no longer predictable—even if, once observed, they could be rationalized—to a situation of extreme complexity, in which the new properties, and thus the deriving functions, can neither be predicted nor rationalized.

| From inanimate matter to life |

It is precisely here that passage from simple supramolecular systems to more complex ones begins the transition from chemistry to biology—that is, a transition from inanimate matter to living matter. As always happens in Nature, however, the transition is not obvious, and is not always easy to describe because of a want of a general consensus on the meaning of *minimal life*.

| The cell |

What we know for certain is that when we get to the cell, even the simplest one—for example, the cell of a bacterium that everyone accords as representing life—we find ourselves confronted with an extremely complex chemical system. We no longer have 3, 10 or 100 molecules, but thousands of billions of molecules associated in highly organized supramolecular structures. This is why scientists have failed, thus far, to transform inanimate matter into living matter spontaneously. In effect, we know nothing about how life originated. Yes, there are hypotheses out there, but none of them have had a clear experimental confirmation, at least for now.

The degree of complexity of the human cell is far greater than that of bacterial cells. First of all, the human cell contains a larger number of components (more than 100 000 billion molecules), and has, above all else, the highest level of organization. Inside the membranes, we find the cytoplasm, organelles and the nucleus. In turn, the latter contains 46 chromosomes, ordered into 23 couples, within which we find deoxyribonucleic acid (DNA). Other than DNA, the most well-known components of the cell—ribonucleic acid (RNA), proteins (amino acid chains), carbohydrates (polysaccharides), lipids (esters of fatty acids) and of course water—are also fundamental for sustaining the life of the cell. To get some idea of the incredible organization and the extreme complexity of the cell, let us consider only the DNA: it is the bearer of genetic information and is present in all the cells, even in the simpler bacterial cells.

| DNA: a complex chemical system |

From a chemical point of view, DNA is a supramolecular system of large dimension and complexity, formed from two long filaments wrapped around one another, forming a double helical structure (Figure 5.3). Every filament is constituted by a sequence of units called mononucleotides, each one consisting of three subunits: a

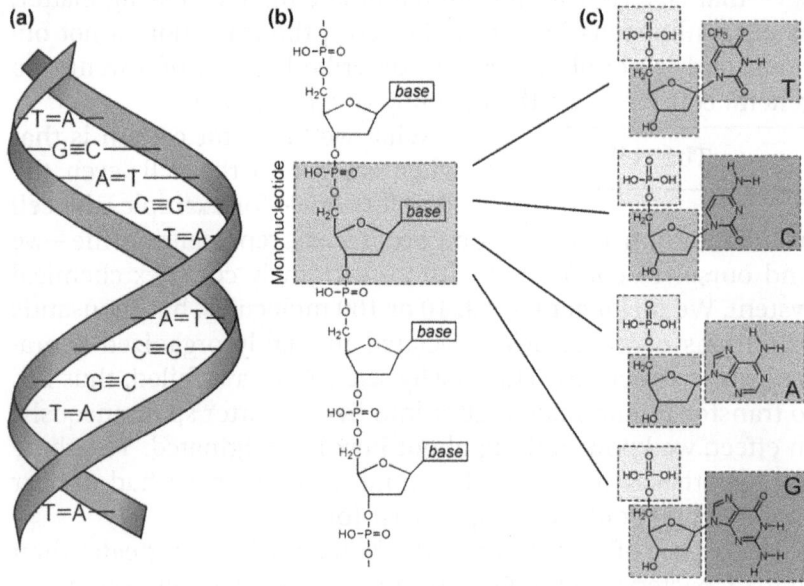

Figure 5.3 Schematic representation of the structure of DNA: (a) segment of the double helix originating from the association of two filaments; (b) segment of a filament in which a mononucleotide is evidenced; (c) the four types of mononucleotides, each of which is constituted by a phosphate group, a sugar and a nitrogenous base that can be of four types, thymine (T), cytosine (C), adenine (A) and guanine (G). The association between the bases cytosine and guanine is shown in detail in Figure 5.2a.

phosphate group, a sugar (deoxyribose) and a nitrogenous base that could be one of four different types: thymine (T), cytosine (C), adenine (A), and guanine (G). The DNA, therefore, is like the scale of a snail whose handrail is made up of the sugar and phosphate moieties and whose steps are formed by two complementary bases, one for each filament. These bases interact through formation of hydrogen bonds, as seen earlier in the case of cytosine and guanine (Figure 5.2a), which are precisely two of the nitrogeneous bases of DNA.

The DNA directs the synthesis of proteins and enzymes necessary for the reproduction of the cells, and for the development of their metabolic activities. This job is carried out by the genes, which are the segments of DNA whose lengths run from about a hundred to a few thousand mononucleotides, ordered according to specific sequences.

Other sequences of mononucleotides, before and after every gene, are predisposed toward the functioning of the same gene; that is, they define when, for how much time and where they must function. Therefore, in order to understand the operation of a gene, it is necessary to know the sequence of mononucleotides that constitute the gene and the interaction with the contiguous segments. This, however, is still not sufficient because the interaction with the other genes, with the other components of the cell, and with the numerous chemical messages that arrive from outside the cell must also be taken into consideration.

The DNA present in a human cell consists of approximately three billion mononucleotides. It can store an amount of information equal to 10^9 bits, similar to that contained in a library of a thousand books of 500 pages each. This aspect is of great importance from the biological point of view, because it is thanks to this *library* that the DNA carries out its function as a bearer of genetic information, written in the sequence of the mononucleotides. Every individual, in fact, has its own specific sequence born with the first cell obtained from the time of conception. In order to form the individual, the number of cells must increase up to 10 000 billion, each one exhibiting the same genetic footprint—that is, the same genome—as the initial cell.

The enzyme DNA polymerase

This arduous task is carried out by an enzyme—the DNA polymerase—that separates the two

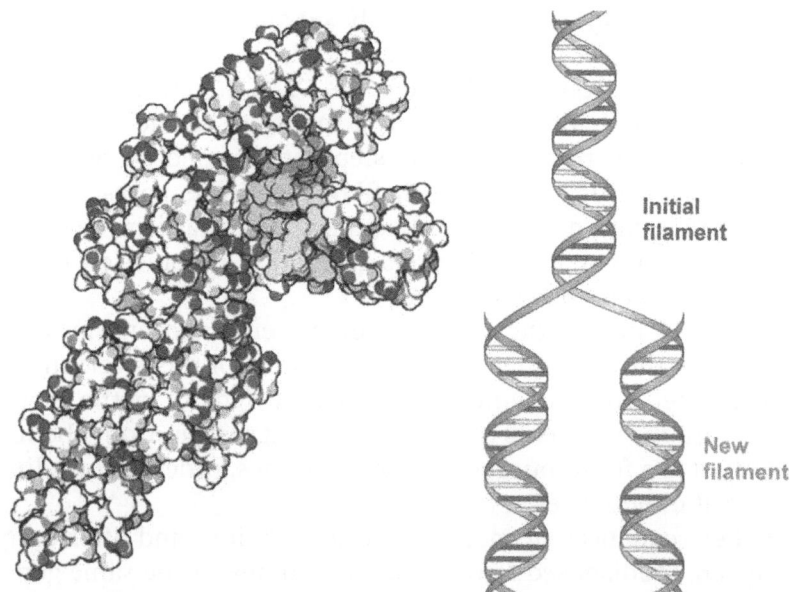

Figure 5.4 Three-dimensional model of the DNA polymerase enzyme (every sphere represents an atom), and a schematic representation of the process of duplication of the DNA to which the enzyme is placed at the head.
Structure on the left was reproduced from http://www.pianetachimica.it/mol_mese/mol_mese_2000/03_DNA_Polimerasi/DNA%20Polimerasi_file/1tau.gif.

filaments and builds the complimentary copy on each, at the end of which two helices are formed identical to the original one. This will constitute the genetic patrimony of two separate cells (Figure 5.4). To understand the efficiency with which this enzyme functions, we need only note that in less than 1 hour it joins the three billion mononucleotides present in the human genome in the exact order and with a precision comparable to that of a typewriter which, in recopying 500 000 pages, makes only one typing error.

This prodigious molecular-level machine works by taking advantage of mechanisms based on molecular recognition. Moreover, it is not the only one present within us. Every cell of our body is, in fact, a most complicated and, at the same time, rigorously ordered and perfectly harmonious ensemble of motors and chemical machines. These have the task to repair the damages that have taken place, to set in action our movements, and to orchestrate our inner world of thoughts, feelings, and emotions.

5.3 FROM CELLS TO MAN

Beyond the cell

The scale of complexity goes well beyond the cell. To obtain a multi-cellular organism, the cells must differentiate and then associate themselves (Figure 5.5) to form the tissues. These, in turn, unite themselves into organs that constitute the several apparatuses whose integration yields the complete organisms. The most complex and complete organism is man, who represents the last step of this scale.

As incredible as it may seem, the complexity of the biological world originates through a spontaneous assemblage of specific chemical compounds that take advantage of the molecular recognition phenomenon, albeit in a gradual, more complex manner. We cannot but be dumbfounded in front of these mechanisms, and still more so when we attempt to examine mental activities and how the brain works. In fact, other than being an organ of an unimaginable complexity, the brain is conditioned by embryonic and fetal development. And not least, it feels the influence of the environment. The brain escapes the determinism of the genome, so that every man/woman is a unique and non-reproducible living being.

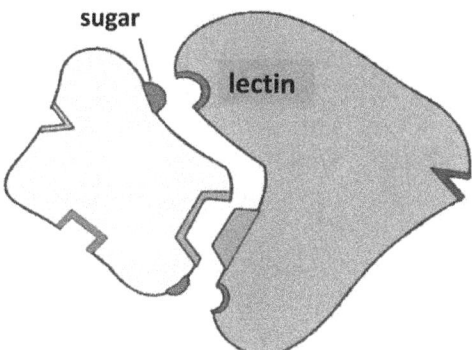

Figure 5.5 Formation of tissues: the cells recognize and associate themselves thanks to the complementarity of chemical substances (sugars and a lectin) that are present on their surface.

5.4 GENETIC ENGINEERING

If at some time we believed that life had a secret and was said vaguely to be the *vital force*, we now realize that the secret of life rests in an extremely complex, diversified and unforeseeable Chemistry that escapes our rational capacity.

> **Intervening on living organisms**

To a large extent, therefore, the *chemistry* of life remains a mystery, even if combined chemical/biological studies have unraveled a considerable number of chemical processes involved in the maintenance and in the natural reproduction of life. These processes allow the chemist to intervene on living organisms and modify their characteristics. With genetic engineering, which is frequently talked about in the mass media, it is possible, for example, to manipulate the DNA through a transfer of genes from one organism to another. Using specific enzymes, it is possible to cleave the DNA at pre-determined points in its sequence and to join, always by means of suitable enzymes, pieces of the DNA originating from different organisms. The DNA so obtained is then said to be a *hybrid* or a *recombinant* DNA. The organisms, plants, and animals whose DNAs have been so manipulated are said to be *transgenic* or *genetically modified organisms* (GMOs).

The usefulness of genetic engineering rests on the fact that the transferred gene continues to carry out its specific function, even in the cell in which it has been transplanted. If, for example, the gene of another organism, whose function was the synthesis of a specific protein, is transferred into the DNA of a bacterium, then because of this genetic modification, the bacterium begins to produce that particular protein, thereby becoming a sort of a living chemical factory. This is the way that insulin and the birth hormone are obtained, which are so useful in the cure of some types of diseases. Genetic engineering, however, can also be used for other uses; for example, to create plants that are resistant to parasites, plants that grow and flourish in barren lands or, better still, to resolve environmental problems.

We must not forget that the use of GMOs can also introduce significant risks. Consider, for example, an environmental emergency such as an oil spill at sea. It might be possible to clean up the mess using a genetically modified bacterium that can destroy the oil at sea. Although the solution may be beneficial, it could expose the aquatic environment to some possible and unpredictable risks, however small, that should never be underestimated. In principle, the bacterium could also begin to proliferate in some uncontrollable fashion and dispose of the oil in other places.

Bioethics

Obviously, the problems become more delicate when humans become directly involved. The mapping of the human genome has opened, among other things, the problem of the genetic record; that is, the classification of individuals based on their genes. Someone has even foreseen that someday we will be able to replace our ID card with a map of our genome. Filing genetic records introduces significant risks that are also not to be underestimated. If, for example, a gene were responsible for a serious disease that manifests itself after many years, the individual bearers of that gene, beyond having their existence ruined to begin with, would have great difficulties in finding a job. Such bearers would likely spend their lives in complete solitude, without the possibility of having a family and children.

Perhaps knowing one's future is therefore not so tantalizing after all. In fact, an article that appeared in 2007 in the magazine *Science* noted, and justifiably so, that the ethical problems connected with genetic records are many and so complex that, if realized, they would be equivalent to opening Pandora's Box. The debate on the feasibility of genetic therapies, the use of embryos, and cloning are topics of much interest currently, and this debate inserts itself into the more general discourse regarding the relationship between ethics, science and technology. In the past few years, the speed of discoveries in the field of biology and the profoundness of their consequences have rendered an ethical appraisal and the definition of rules of behavior quite difficult. It is therefore necessary and urgent that bioethical problems be confronted, keeping in mind the complex material structure and spiritual nature of man.

5.5 BEYOND THE SCALE OF COMPLEXITY

Through a unique and fascinating logic thread, the chemical approach allows for the unification of the invisible world of chemistry to the macroscopic world of biology, which culminates with man. It must not be forgotten, however, that life cannot be reduced to its chemical and biological aspects, just as it is not possible to reduce one of Beethoven's great symphonies to its musical score.

> **Life is more than just chemistry and biology**

To interpret the behavior of man simply on chemical and biological aspects would mean ignoring the more precious assets that distinguish man from all other living beings: his dignity, his freedom of thought, his conscience and the knowledge of his abilities and his limits. There are, in fact, questions that science will never be able to answer. For instance, who made up the laws of Nature? Why are we in this world? What sense does the life of man have in a world of objects? Inasmuch as science cannot answer these questions, man searches for answers in other areas.

Part Two
Chemistry: Yesterday, Today and Tomorrow

Part Two

CHAPTER 6

Reading and Writing with Molecules

6.1 THE CHEMIST: EXPLORER AND INVENTOR

For many years, the role of chemistry has essentially been to establish the composition and the structure of natural products, and to understand the rules that Nature imposes on chemical phenomena. The situation, however, has changed with time.

> **Natural versus artificial**

In the operation *acquaintance of Nature* and on the revelation of the secrets of natural processes, the chemist has become so skillful and such an expert as to succeed in synthesizing compounds and in finding processes that do not exist in Nature. Such compounds have become known as *artificial products*. In the common language, the word *artificial* is often used in a disdainful sense, but in fact its meaning is quite different because it indicates *a product from the intelligence of man*.

Chemistry: Reading and Writing the Book of Nature
By Vincenzo Balzani and Margherita Venturi
Translation by Nick Serpone
© The Royal Society of Chemistry 2014
Published by the Royal Society of Chemistry, www.rsc.org

As the explorer of Nature more often than not today's chemist is being kept abreast by the chemist inventor and by the chemist engineer at the atomic and molecular levels. Indeed, chemistry is a *book* not only *to be read* (natural substances and processes), but also *to be written* (artificial substances and processes). Moreover, the progress made in the last few years has clearly demonstrated that, if the part not yet read is immense, then the part yet to be written is practically infinite.

6.2 THE MERRY-GO-ROUND OF CURIOSITY

Scientists and artists in front of a tree

The scientist is in love with his job, which is often compared to that of an artist, even if, in fact, there are significant differences between the job of a scientist and that of an artist. They reflect different mentalities. For example, in front of a magnificent tree, like a great oak, a poet will stop, contemplate and then may write a poem (Figure 6.1).

Figure 6.1 The poem *Trees* by Frank S. Flint.

A great writer can put pen to paper and write a memorable piece:

On the edge of the road there was an oak, which was probably ten times older than the birches of the forest, ten times larger and two times taller than any birch. It was an immense oak.
                                         War and Peace, Lev N. Tolstoy

An artist like Piet Mondrian can draw inspiration to paint a magnificent picture, *The Red Tree* (the black and white image below does not do justice to the beauty of this picture):

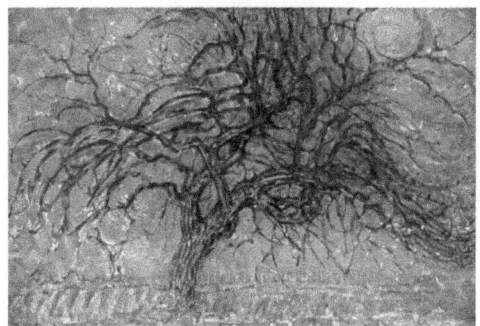

Even the scientist remains fascinated in seeing a great oak, but his curiosity and his knowledge push him to go beyond the aesthetic beauty.

The first thing the scientist thinks about is that the tree exists because there is air and the Sun; indeed, he knows that the oak is precisely air and Sun, as admirably stated by the 1965 Nobel prize winner for physics, Richard P. Feynman:

> *A tree is essentially made of air and Sun.*
> *When it is burned, it goes back to air,*
> *and in the flaming heat is released*
> *the flaming heat of the Sun*
> *which was bound in*
> *to convert the air into tree*

The scientist asks himself questions that the artist is not interested in; for example, why is sunlight necessary to grow a tree? How does a tree use the sunlight to produce its fruits? What is sunlight?

> **The carousel of questions and answers triggered by curiosity**

The scientist is a person who knows that he does not know everything; he is a curious person, very curious indeed. Poor and astonished in front of the intricacy and the beauty of the world around him, the scientist asks questions to satisfy his curiosity, which he then addresses to Nature in the form of experiments (Figure 6.2). Naturally, they must be intelligent questions, that is, experiments devised with imagination, prepared with care and executed with rigor. The more intelligent the question, the more important will be the answer that he gets back from Nature. Some scientists are of the opinion that the great discoveries of science will come from answers to questions that we are still not in a position to formulate.

Having done the experiment, the scientist now listens to what Nature wants to communicate to him. He listens passionately. And as Albert Szent-Gyorgyi noted: *Discovery consists of seeing what everybody has seen and thinking what nobody has thought.*

Listening to the answers that Nature gives to his questions, the scientist learns, acquires new knowledge and feels some

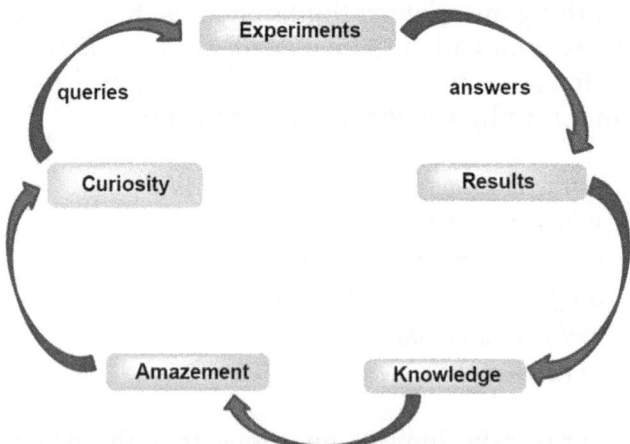

Figure 6.2 The merry-go-round of curiosity. Every scientific discovery generates more questions than answers.

astonishment that generates further curiosity in him. In turn, this encourages new experiments that bear new results, new understanding, and new astonishment. One would think that this merry-go-round of questions and answers would eventually come to an end. But this does not happen, because every discovery generates more questions than answers.

6.3 NATURAL MOLECULES

Reading the book of Nature

In Nature—that is, in the ground, the seas, the air and in vegetable and animal organisms—one can find a large variety of molecules. Some natural molecules are very simple: the oxygen molecule, O_2; the molecule of carbon monoxide, CO; the water molecule, H_2O. Others are of medium complexity: the molecule of acetic acid, $C_2H_4O_2$; the glucose molecule, $C_6H_{12}O_6$. Other molecules are very complex such as, for example, insulin—the hormone produced by the pancreas for regulating the level of glucose in the blood—which is constituted of 778 atoms; the molecular formula is $C_{254}H_{377}N_{65}O_{76}S_6$.

Many natural molecules, presumably the more abundant and more important ones, have been identified and investigated in every detail. These investigations have clarified which and how many atoms make up these molecules, have established how the atoms are bonded, have determined the three-dimensional structures of the molecules, and have investigated their properties—for example, their solubility, their melting point, the internal movement of their atoms and their reactivity, among many others.

The spiciest molecule is *capsaicin* present in *peperoncino* (hot pepper):

Countless other natural molecules have been characterized, but their properties have yet to be fully investigated. Lastly, an unknown number of other natural molecules, most likely a large number, have yet to be identified. Taken all together, the natural molecules are part of the great book of Nature, together with the plants, the animals and the minerals. As in other cases, even the world of molecules has been discovered only partially and only a small fraction of the book has been read more or less carefully.

Many natural molecules, even complicated ones, after having been read in every detail, have subsequently been synthesized (rewritten) in a laboratory, beginning with simple compounds.

6.4 HOW WE GOT TO *ASPIRIN*

To understand how we go from reading to writing chemistry, let's consider, for example, the case of salicylic acid, a substance from which we can synthesize the most famous of known drugs: *Aspirin*.

> Learning to create new compounds by reading the book of Nature

Substances extracted from the bark of some plants and used as pain-killers were well known hundreds of years ago. In 1763, the English Reverend Edward Stone introduced for the first time—in a scientific way—the antipyretic properties of the extracts from the bark of the willow tree. In 1828, after treating the bark with water, followed by filtration of the insoluble parts and evaporation of the solution, the German pharmacologist Johann Andreas Buchner obtained a yellow substance that he named *salicin*. Toward 1830, the Italian chemist Raffaele Piria separated a colorless crystalline substance from a salicin solution that had acidic properties, and so called it *salicylic acid*. He also recognized that this substance was responsible for the antipyretic action. In 1859, the German chemist Kolbe discovered that salicylic acid could be decomposed into phenol and carbon dioxide (Figure 6.3), thereby succeeding in describing its chemical formula.

Once the formula was known, it was possible to reproduce salicylic acid in the laboratory (1874), thus making it possible to commercialize it. However, it was soon discovered that its beneficial effects were outweighed by significant disadvantages: an unpleasant taste and irritations to the stomach, among others.

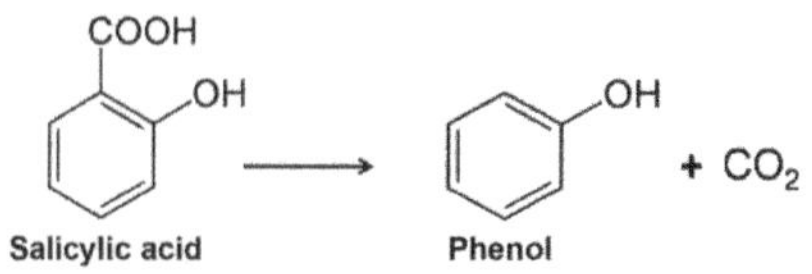

Salicylic acid → Phenol + CO_2

Figure 6.3 Reaction illustrating the decomposition of salicylic acid.

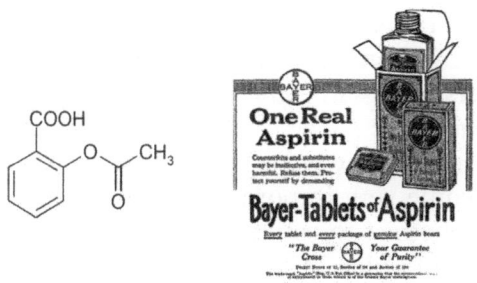

Figure 6.4 Structural formula of *acetylsalicylic acid* (ASA) and one of the first of many advertisements of Aspirin.
Reproduced from: http://www.in-farmacia.it/files/84-xcms_label_small220px-Bayer_Aspirin_ad,_NYT,_February_19,_1917.jpg.

Chemists then began to prepare—to synthesize is a more appropriate chemical terminology—several derivatives of salicylic acid until they discovered that the molecule of *acetylsalicylic acid* (Figure 6.4) displayed only minor side effects, but retained the beneficial properties of salicylic acid. The synthesis of this new molecule, which was different from the molecule already existing in Nature, but which was inspired by it, led to the initial stages of the modern drug industry. In 1899 in fact, Bayer, a giant of the German chemical industry, patented the drug that ultimately came to be known as *Aspirin* (Figure 6.4), and soon began to produce and commercialize it in very large quantities.

From then on, Aspirin became the drug that has contributed the most to the well-being of man. So much so that the agency responsible for health-related issues in the USA—the Food and Drug Administration (FDA)—recommends that people 50 to 80 years old take a daily low dose of aspirin (81 mg) to prevent heart attacks and other cardiovascular diseases. Not only does Aspirin have beneficial effects for well-being, but it also has significant economic benefits to both the individual and society.

At the present time, reading the book of natural substances is very fast, thanks to remarkable progress in the techniques of separation and characterization of chemical compounds. For example, the identification of the active ingredient of a natural substance can be carried out within a few months. The same can be said for research on derivatives of a drug so as to optimize its beneficial effects. For this purpose, a novel approach that has proven most effective and useful is *combinatorial chemistry*, by means of which thousands of chemical substances can be synthesized and simultaneously experimented with for their biochemical effects. The same methodology is being exploited in *materials chemistry*.

For the planning of new drugs, studies that involve the computer have become very important (*computational chemistry*) as they allow the chemist to estimate the optimal dimensions and the shapes of molecules in relation to their end functions; for example, interaction with a specific protein. However, on the other side of the coin is that the time needed to bring a new drug to the market today is long. This is due to the demands of detailed experimental and precautionary investigations that ensure not only the effectiveness of the drug, but also the total absence of toxicity and side effects that could be harmful to man.

6.5 ARTIFICIAL MOLECULES

> **Where Nature ends, man begins**

Chemists have already added tens of millions of artificial molecules to the large number of molecules that exist in Nature. Thanks to the profound understanding of several types of chemical reactions, it is now possible to prepare molecules of whatever shape and dimension one wishes.

Thus, a famous sentence said by Leonardo da Vinci is also true in the case of chemistry: *Where Nature ends to produce its species, man begins to create an infinite number of species in harmony with the laws of Nature.*

> **Synthesize new molecules, but with care!**

To produce artificial molecules, it is necessary to have at one's disposal the proper *construction materials* at the molecular level: that is, other

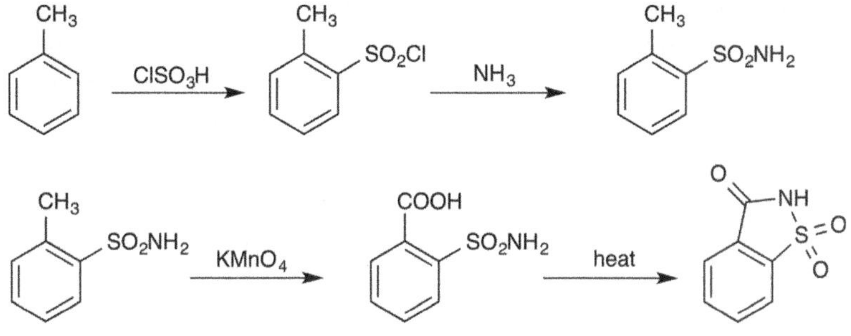

Reading and Writing with Molecules 75

Figure 6.5 Synthesis of saccharin starting from a derivative in the distillation of crude oil: toluene.

molecules utilizable as raw materials. Chemical synthesis can, in fact, be considered, as we said earlier, a sort of a game of Lego. The construction material to synthesize drugs, dyes, pesticides, plastics, *etc.* is obtained in large measure, directly or indirectly, from natural gas (mostly methane) or from petroleum (crude oil) that consist of a complex mixture of many chemical products, for the most part hydrocarbons. Some of these compounds are used to construct more complex molecules, while others are fragmented (the cracking process) so as to obtain smaller molecules from which the desired molecules can be synthesized.

The synthesis of a new molecule or a new material is a complex process that requires ideas, organization and work. For example, Figure 6.5 illustrates the synthesis of saccharin, a famous artificial sweetener, beginning from methylbenzene (also known as toluene), a derivative of the distillation of petroleum.

Saccharin has a greater sweetening power—approximately 450 times—than sucrose (sugar) and, like other artificial sweeteners (aspartame, cyclamate), is used instead of sugar by diabetic people, or otherwise in slimming diets.

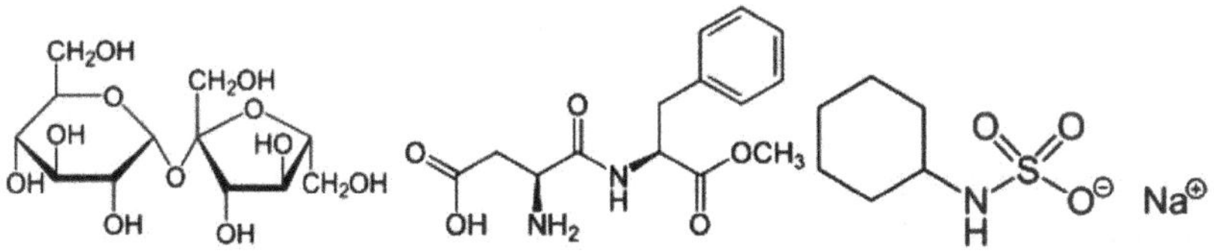

Sucrose   **Aspartame**   **Cyclamate**

To obtain artificial compounds, successive chemical reactions are often required (4 in the case of saccharin), each of which must be thought of and planned to the last detail: reagents, reaction environment (solvent), temperature, catalysts and when necessary use of light or some other form of radiation, so as to trigger or accelerate the desired reaction, as well as appropriate analytical techniques to establish whether the reaction has occurred or not.

> The most sweetening molecule is sucronic acid:
>
> Its sweetening power is 300 times greater than that of saccharin (the artificial sweetener used by diabetics).

Chemical processes always have a certain degree of risks, because they often require high temperatures and pressures. The chemical industry, therefore, must adopt particular preventative measures to avoid incidents.

As in every human endeavor, even chemical syntheses can be used for either constructive or destructive purposes: for example, a pesticide can be synthesized in order to protect agricultural harvests, or else can be used as a deadly poison to use against one's enemy (chemical weapons). Indeed, in many cases the primary reagents or intermediate products to produce a useful pesticide or a detrimental chemical weapon are the same. As such, industries with no scruples whatsoever can easily hide the real purpose of their production. It must be emphasized that the use of an artificial product—even if it is the fruit of research undertaken with the best of intentions—may unavoidably involve a certain margin of risk and, therefore, must always be used with great care. This is particularly true in the case of drugs.

On 10 July 1976 there was an explosion of a chemical reactor at the ICMESA Company in the municipality of Seveso (Italy). It provoked spillage of a cloud of dioxin, one of the most dangerous of toxic substances:

The toxic cloud covered an immense land area of the Brianza lowlands. Although there were no fatalities, there were serious consequences on the health of the population, which remains to this day a subject of serious studies. Vegetables covered by the toxic cloud dried up because of the highly herbicidal power of the dioxin; thousands of animals were contaminated and had to be killed.

CHAPTER 7

Creativity and Beauty

7.1 FOR BETTER OR FOR WORSE

Creativity and beauty are often considered to belong exclusively to the world of art. In fact, creativity and beauty are also two characteristics of science, in general, and of chemistry, in particular.

Chemists: from explorers to inventors

The creativity of chemists began when, as explorers of Nature, they became inventors—that is, when they began to synthesize molecules in the laboratory that did not exist in Nature. A significant part of their creative work has produced important benefits to society. Molecules of drugs were invented that were capable of curing many diseases, or else of making us feel no pain (the anesthetics). Also invented were molecules that protect us from the cold and the heat (thermal insulators), molecules that help the growth of robust harvests (fertilizers) and protect them from parasites (fungicides); molecules to color textiles and objects that are used daily (dyes), molecules that make foods and drinks more delicious (additives), molecules that protect the eyes from intense light (photochromics) and protect the skin when exposed

---

Chemistry: Reading and Writing the Book of Nature
By Vincenzo Balzani and Margherita Venturi
Translation by Nick Serpone
© The Royal Society of Chemistry 2014
Published by the Royal Society of Chemistry, www.rsc.org

Creativity and Beauty

to solar radiation (sunscreens, solar creams), molecules that protect us from insect bites (repellants), and molecules that project a fragrance more pleasant than molecules produced by flowers (artificial scents, perfumes).

Chemists have also created beautiful yet intriguing molecules in the shape of a tree, a knot, a chain, a bridge or a dome. They have learned to move up in the scale of complexity. Working as engineers and architects, they have used molecules to create molecular devices and machines with nanometer dimensions capable of performing, in some cases, more intelligent and complex functions than those invented by Nature.

Below are the structures of the most explosive molecules:

NG   TNT   RDX   HMX

Nitroglycerin (NG) was discovered by the Italian chemist Ascanio Sobrero in 1847; Alfred Nobel later discovered that when stirred with fossil flour it forms a more stable explosive, which he later patented as dynamite. This was the first commercial explosive (1870), which was followed in 1910 by trinitrotoluene (TNT). Cyclictrimethylene-trinitroamine (RDX) is the more economic explosive, while cyclictetramethylene-tetranitroamine (HMX), commercialized in 1955, is the most powerful explosive.

Unfortunately, as always happens in science, creativity can also follow misguided paths and lead to monstrous creatures capable of mass destruction. Regrettably, chemists are not free from this sin, as they have also created molecules that can destroy (explosives), cause devastating fires (incendiary bombs), kill people (poisons) and annihilate entire populations (chemical weapons).

7.2 BEAUTIFUL MOLECULES

In more recent years, chemists have been busy synthesizing molecules of every shape and size. The extraordinary wealth of the molecular world created by chemists can hardly be subjected to an organized nomenclature. For this reason and because of the more diffuse understanding of what molecules are—real *objects*, even though of nanometer dimensions—there is a tendency to abandon the official nomenclature, which by now is insupportably difficult particularly in naming complex molecules. Accordingly, chemists have tended to use fancy names borrowed from objects encountered daily. This is so true in fact that even in articles published in most reputable scientific journals, we cannot but take note of molecules that possess the shape of a pincer, a butterfly, a bridge, a box, a collar, a gondola, a thread, a chain, a barrel, a basket, a belt, a cup, a cage, a chain, a crown, a crypt, a fence, a soccer ball, a door, a net, a propeller, a hinge, a knot, a ladder, a spider, a polyp, an egg, a pagoda, a shelf, a wheel, a scorpion and a star, among others.

> **It happens in chemistry as in architecture**

In many cases, the molecules are highly symmetric systems and possess fascinating shapes. In effect, it is possible to have some idea of these peculiarities by observing their graphical illustrations in Figure 7.1. As was justly written by Primo Levi in his book *The Periodic System*:

> *In fact, it also happens in chemistry, as in architecture, that the most beautiful buildings, which are simple and symmetrical, are also the more balanced; in fact, it also happens for molecules as for the domes of cathedrals and the arches of bridges.*

Some molecular structures are so esthetically enjoyable that they can be taken as models to create beautiful sculptures (Figure 7.2).

Creativity and Beauty

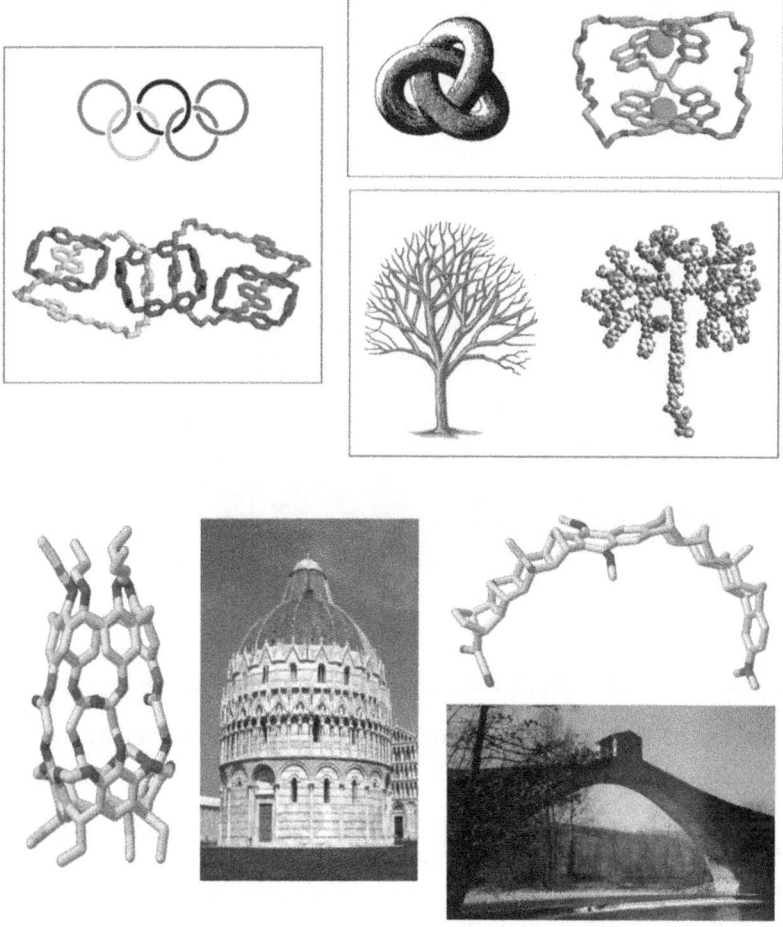

Figure 7.1 Examples of beautiful molecules synthesized by chemists that, by their shape, recall objects encountered daily and architectural structures.

7.3 CHEMISTRY IN THE WORDS OF SCIENTISTS AND WRITERS

The language and the concepts of chemistry have often been arranged in an admirable fashion by scientists and writers. A splendid example is the description of a tree given by Richard P. Feynman (see Chapter 6).

> **The profession of a chemist**

It is difficult to explain in layman's terms the beauty and the intricacy of chemistry, and the work of

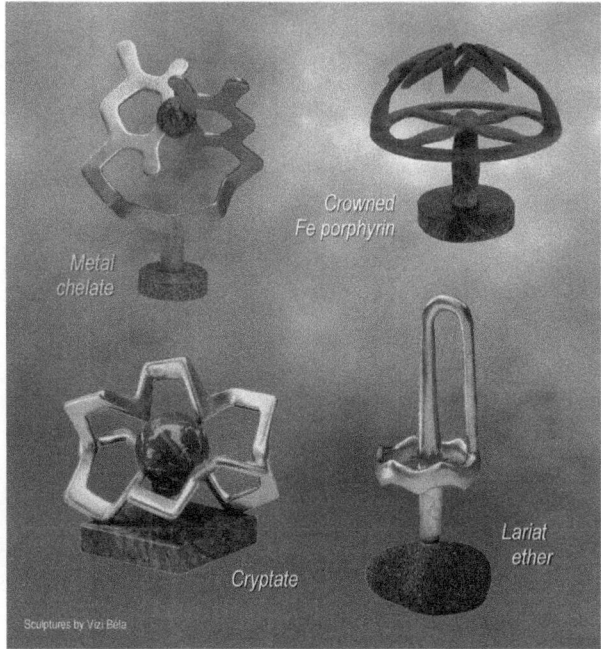

Figure 7.2 Four molecular structures represented in the sculptures of the Hungarian artist Béla Vízi.
Reproduced from http://www.mi.sanu.ac.rs/vismath/visbook/vizi/index.html.

chemists. However, Primo Levi, a great chemist and author, succeeded effortlessly. In his already mentioned book *The Monkey's Wrench*, he envisioned speaking to a mechanic and succeeded in describing, somewhat poetically, the profession of a chemist.

> *My profession, my real one, the profession I studied in school and that has kept me alive so far is the profession of chemist. I don't know if you have a clear idea of it, but it is a bit like yours; only we rig and dismantle very tiny constructions. We are divided into two main branches, those who rig and those who dismantle or break down, and both kinds are like blind people with sensitive fingers ... so we have invented various intelligent gadgets to recognize them (molecules) without seeing them ... Those who dismantle, the analytical chemists,*

Creativity and Beauty

in other words, have to be able to take a structure apart piece by piece without damaging it, or at least without damaging it too much; then they have to line up the pieces on the desk, naturally without ever seeing them, but recognizing them one by one. Then, they say in what order the pieces were attached.

Levi also described the chemists that assemble—that is, those that synthesize made-to-measure chemical systems. In another of his books cited earlier, *The Periodic System*, this chemist-author described, again in a somewhat poetic and fascinating manner, the voyage of a carbon atom with no end in sight—initially located in limestone (calcium carbonate, $CaCO_3$)—that when burned in a furnace ends up as a molecule of carbon dioxide (CO_2):

... firmly clinging to two oxygen companions, it issued from the chimney and took the path of the air ... It was caught by the wind, flung down on the earth, lifted ten kilometers high. It was breathed in by a falcon, descending into its precipitous lungs, but did not penetrate its rich blood and was expelled. It dissolved three times in the water of the sea, one in the water of a cascading torrent, and again was expelled. It traveled with the wind for eight years: now high, now low, on the sea and among clouds, over forests, deserts, and limitless expanses of ice; then it stumbled into capture and the organic adventure ... Our atom of carbon enters the leaf, colliding with other innumerable molecules of nitrogen and oxygen. It adheres to a large and complicated molecule that activates it, and simultaneously receives the decisive message from the sky, in the flashing form of a packed of solar light: in an instant, like an insect caught by a spider, it is separated from its oxygen, combined with hydrogen and finally inserted in a chain, whether long or short does not matter, but is the chain of life.

This likeable story emphasizes the strict connection that exists between all shapes of matter present on Earth.

| The climbing vine |

Chemistry, words and life have been integrated splendidly by Roald

Hoffmann—a poet, a philosopher and winner of the 1981 Nobel Prize for Chemistry. In a work titled *Sustainable Development*, Hoffmann emphasizes the miraculous power of photosynthesis, describing the unstoppable entanglement of a climbing vine around a big tree. Here are some of the first few verses:

> *Alive? The vines just push the question aside,*
> *A green muff for these trees,*
> *coating them real tight*
> *like a cross-linked polymer gone mad.*
> *The problem in spring is the trees'- are they?*
> *And will they be? ...*

7.4 INTELLIGENT MOLECULES

The most creative accomplishment of chemistry has always been considered the planning and the synthesis of new molecules. In several fields of arts and science, however, creativity often manifests itself, not through new things, but by an original use of previously known materials or patterns.

Likewise, creativity in chemistry has been manifested through new conceptual interpretations of classes of reactions that had been examined previously. A typical example is the discovery that fundamental chemical processes (acid–base reactions, oxidation–reduction (redox) reactions and isomerization reactions of well-known molecules) can be used to achieve logical functions.

In this regard, we should first of all recall that molecules can be *read* through numerous chemical and physical analytical methods that underlie analytical chemistry. In general, to understand what type of molecules is present in a sample, we send a signal to the sample. If the molecules of a certain type are present, they respond with a signal whose intensity is related to the number of such molecules.

We can also *write* with molecules. This occurs every time that a molecule involved in a chemical reaction is transformed into another. We can also write on molecules using light (photochemical reactions) and electricity (electrochemical reactions).

Creativity and Beauty

With analytical chemistry, we can read what exists before and after the reaction has occurred, so that it is possible to understand what and how much has been written.

Molecules for remembering

Molecules can also be used to construct memory, that is, *to remember*. Molecules that are suitable in performing such a function are those that are capable of existing in two forms: A and B (isomers), each with different properties from the other—for example, color—and one can be converted into the other by means of appropriate external stimuli (triggers). Examples of this type of molecules are provided by compounds that have a –C=C– double bond. Accordingly, starting from the more stable form A, we can write on it with an external signal (for example, a photon) that converts A into the less stable form B. This operation is equivalent to writing and storing a bite of information. The B form can return, more or less rapidly, back to its initial A form (reaction 1 of Figure 7.3), or else can remain as is (reaction 2). The first case corresponds to a labile memory (the written information is cancelled with time), whereas the second case corresponds to a permanent memory. In the second case, it is possible to force B to return back to A by means of another external stimulus (for instance, a photon of a different frequency; reaction 3). This is the same as canceling the information written previously.

$$A \xrightleftharpoons{h\nu} B \quad (1)$$

$$A \xrightarrow{h\nu} B \quad (2)$$

$$A \xrightleftharpoons[h\nu']{h\nu} B \quad (3)$$

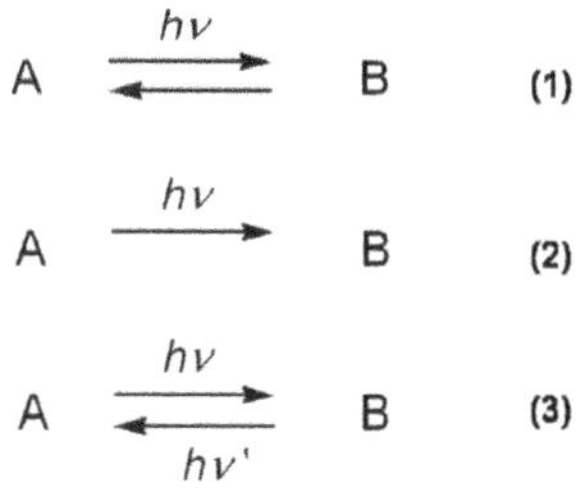

Figure 7.3 When a molecule can exist in two interconvertible forms, A and B, it can function either as a labile memory (1) or as a permanent memory (2). In this second case, we can intervene through another external stimulus to cancel the information written previously (3); $h\nu$ and $h\nu'$ represent the luminous stimuli (the photons).

With some particular molecules, the operations *reading, writing* and *remembering* can be combined and elaborated to effect logical operations used in information processes.

Logic gates

To understand what a logic operation is, consider an electrical circuit consisting of a lamp, a battery and 2 switches connected in series (Figure 7.4). The light is lit only when both switches are closed (logic AND). If the switches are connected in parallel—rather than in series—then to switch on the light suffices to close one or the other switch (logic OR). There exist other more complex logics, one of which consists of switches connected in such a way that the light is switched on *only* if one of the two switches is closed and not both (logic XOR). These and other logic operations can be performed by using molecules and signals that can read and write on such molecules.

A molecular level AND logic gate

For example, the system shown in Figure 7.5 behaves in solution as the AND logic gate. The central portion

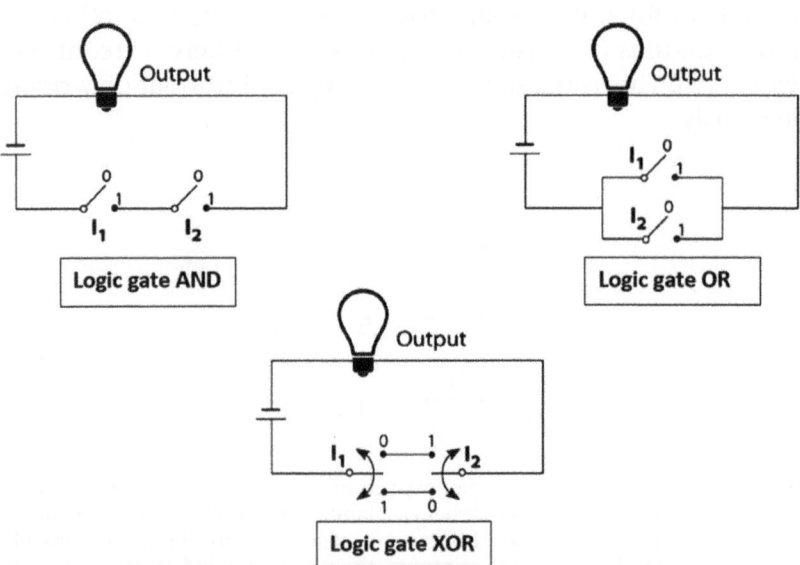

Figure 7.4 Some examples of logic gates and their representation as electrical circuits.

Creativity and Beauty

of this system is a molecule of anthracene (*1*), a chemical species that when irradiated by UV light emits visible light (a phenomenon known as *fluorescence*). When anthracene is connected to an amine group (*2*) and to a crown ether group (*3*), excitation with UV light does not induce fluorescence from the anthracene unit because the contiguous amine and crown ether groups quench (that is, extinguish) the fluorescence. However, if protons (H^+) and sodium ions (Na^+) are added to the solution, they associate with the amine and crown ether units, respectively, and modify their properties, thereby preventing the quenching of the anthracene unit.

Thus, addition of H^+ and Na^+ has the same effect as closing the two switches in series in the electrical circuit AND of Figure 7.4. Thanks to these principles, a molecule can also be used to add signals—that is, to perform calculations—and to perform more complex logic functions such as, for example, encoding and decoding information.

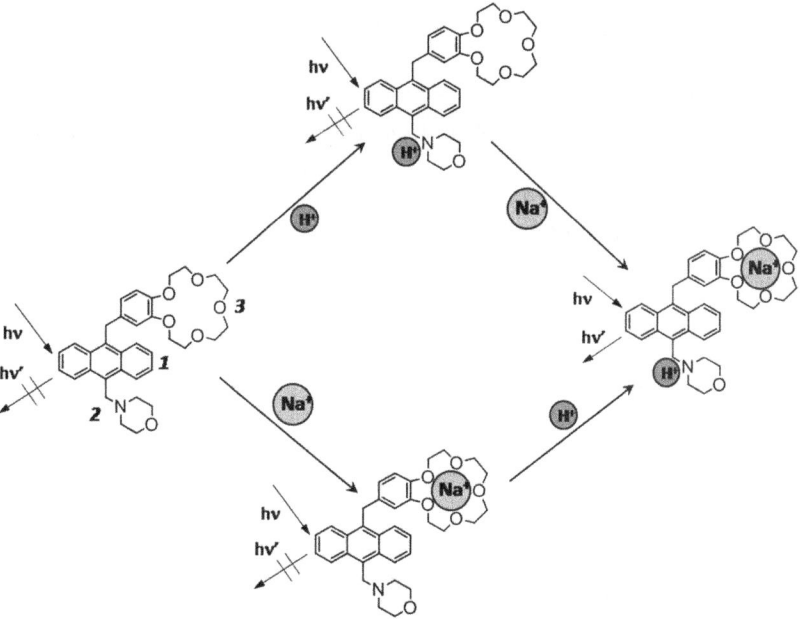

Figure 7.5 A supramolecular system that behaves as a logic gate AND. The arrows with the suffix h*v* indicate absorption of light, while those with h*v'* indicate emission of light. The double bar lines indicate the absence of emission.

7.5 MOLECULAR MACHINES

In the last few years chemists have learned to work as molecular engineers by assembling specific molecular units to obtain supramolecular systems capable of performing functions more valuable than those performed by separated molecules. Roald Hoffmann best defined this work as: *The marriage between the synthetic talent of chemists and the engineering mentality.*

| Molecular engineering |

The pioneer of supramolecular chemistry—the branch of chemistry that forms the base of molecular engineering—is the 1998 Nobel Prize for Chemistry winner, Jean-Marie Lehn. Note, however, that the notions of molecular engineering and supramolecular chemistry were articulated long ago by Primo Levi in his book already mentioned in this chapter:

> *It is reasonable to proceed a bit at a time, first attaching two pieces, then adding a third, and so on ... we don't have those tweezers we often dream of at night, the way a thirsty man dreams of springs, that would allow us to pick up a segment, hold it firm and straight, and paste it in the right direction on the segment that has already been assembled. If we had those tweezers (and it's possible that, one day, we will), we would have managed to create some lovely things that so far only the Almighty has made, for example, to assemble—perhaps not a frog or a dragonfly—but at least a microbe or the spore of a mold.*

As of now, no one has succeeded in assembling a frog, a dragonfly or even a microbe. However, using the concepts of molecular engineering, chemists have succeeded in constructing molecular-level machines; that is, supramolecular systems capable of performing mechanical movements on command.

| A four-stroke molecular engine |

It might seem strange to talk about artificial molecular machines, but we should remember that all living organisms contain natural molecular machines that make them move, speak and see. The cells in our body have hundreds of different types of

Creativity and Beauty

molecular machines; each one is specialized in performing a specific function. Natural molecular machines are perfect, phenomenal, but also very complex. The ones that chemists create in a laboratory environment are obviously far more simple, but nonetheless fascinating.

An example of a supramolecular system that acts as a molecular machine is illustrated in Figure 7.6. It is a rotaxane—so-called because it has the structure of a *wheel* threaded on an *axle*. It was planned in such a way that every light impulse absorbed by the $[Ru(bpy)_3]^{2+}$ unit (the first one on the left hand side of Figure 7.6) causes an alternating movement of the ring through four stages between two well-defined positions along the wire. At the molecular level, the behavior parallels that of a four-stroke linear engine activated by light.

One could easily question the usefulness of these systems. At the moment, we have no answer because such systems were constructed as part of fundamental studies carried out to

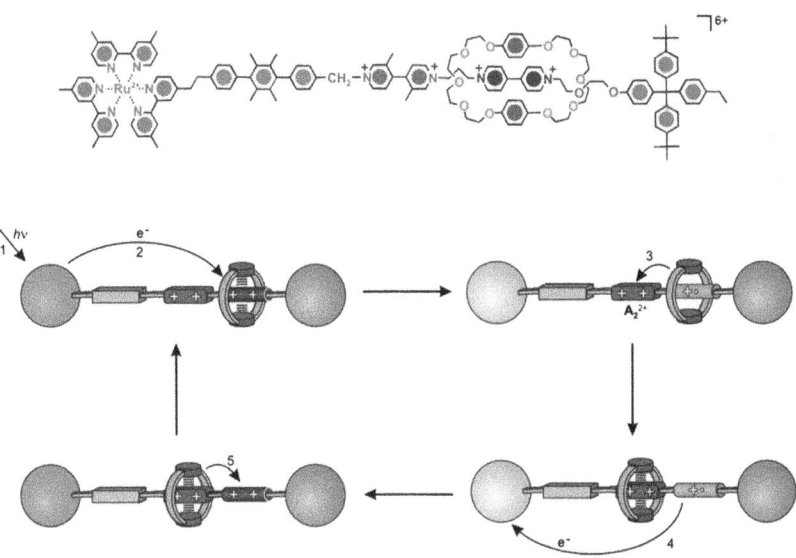

Figure 7.6 Chemical formula of a supramolecular system (length: 6 billionth of a meter; 6 nm) and schematic representation of the sequence of ring movements (3 and 5) induced by absorption of a photon by the unit $[Ru(bpy)_3]^{2+}$ (1) and by the subsequent electron transfer processes (2 and 4).

advance scientific knowledge. We could, however, hide behind a famous anecdote from an encounter between Faraday and the then English Prime Minister Gladstone, who wished to know the purpose of the esoteric substance *electricity* discovered by the scientist, to which Faraday responded: *One day Sir, you may tax it;* a lapidary response perhaps, yet at the same time forward-looking and typical of someone who was well aware that science is full of surprises.

One thing can be said without fear of being contradicted. The planning and the construction of molecular devices and molecular machines demonstrate how relevant creativity is in the development of chemistry. We have already emphasized that chemistry is comparable to a book that is not only *to be read*, but also *to be written*. Only people who commit their pen to creativity, without forgetting the great responsibility that comes with it, will be able to write new and important pages in this book.

Part Three
Teaching and Science

CHAPTER 8

On Teaching: What and How

8.1 CHEMISTRY AS A SCHOOL SUBJECT

Textbooks, encyclopedias and dictionaries define chemistry as: the science that investigates the properties, the structure, the preparation and the reactions of chemical elements and of their compounds. However, though correct and to the point, this definition does not describe the immediate importance, the usefulness and, above all, the beauty of chemistry.

| **Chemistry is not an arid subject** |

It is appalling that these aspects, as well as not appearing in the definition of chemistry, are also absent from the majority of chemistry courses, so that this discipline is being experienced by students as something mysterious and is perceived as having nothing to do with daily activities. Every year, thousands of students are forced to take some kind of chemistry course, and have to study from somewhat arcane and sometimes boring textbooks. We should therefore not be surprised if the few chemical notions that students retain often consist of memorized formulas and some fictitious phrases regurgitated *ad nauseam* at examination time. Too often they spurt out concepts so distorted as to be

---

Chemistry: Reading and Writing the Book of Nature
By Vincenzo Balzani and Margherita Venturi
Translation by Nick Serpone
© The Royal Society of Chemistry 2014
Published by the Royal Society of Chemistry, www.rsc.org

unrecognizable. Hence, we should not be shocked if journalists, politicians, administrators, and even cultured people express inappropriate and too often damaging comments about chemistry.

This problem has led to a considerable drop in undergraduate registrations for a major in the chemical sciences. And to think that right now there is a great demand for chemistry graduates who are proud of their education and conscious of their fundamental role in society. Then, there is also the consideration that a good fundamental chemical know-how would be very useful for laypeople to have, if only to allow them to make personal choices, to be aware and motivated, and to express thoughtful and coherent opinions on certain issues of great social importance, such as environmental pollution, energy resources and global warming, among others.

It is necessary then to attract students back to chemistry. To achieve this, we provide some suggestions on how to do so in the following two sections. These should prove useful to those who teach chemistry. In this regard, the influential scientific magazine *Science* dedicated two entire issues on this subject, which were published on April 23, 2010 and April 19, 2013. Three methodologies emerged to allow students to regain the language of chemistry and its operational approaches. These were:

(1) confront issues associated with the daily realities and in the social context;
(2) utilize an interdisciplinary approach;
(3) exploit hands-on inquiry-based learning.

8.2 WHAT TO TEACH

An outline of a teaching scheme is proposed herein as a follow-up which, based on what was said above, involves, among others, the necessity to renew the programs, limit the contents, and follow a precise and logical course of action that brings forth the interdisciplinary nature of chemistry. We wish to emphasize that the teacher should not be preoccupied if, during the lectures, he is unable to answer all students' questions exhaustively and conclusively. He should, in fact, consider this limitation a positive aspect of his teaching, as it is useful to stimulate the students' imagination and curiosity, and force them to search for more rigorous explanations at a more advanced level. Indeed, as the Greek philosopher and botanist Theophrastus (*ca.* 371–287 BC) wrote: *Education is not filling a bucket, but lighting a fire.*

Another consideration regards the inevitable fact that there may be other disciplines that may be perceived as more interesting by students, just as there may be those perceived as less interesting. This can occur in all fields of study and at any moment in the students' life. It is entirely appropriate for students to express certain preferences and continually make certain choices. Accordingly, it is important to tell students that it is good to have preferences, as it is an essential contribution to the formation of their personality, a result of a continuous selection process. It is also equally important that students understand the need to study all disciplines with determination and personal commitment, even those they may not like. Someday in the not too distant future, they may remember them and come to like them, or else find the necessity to use them.

8.2.1 First Thematic Nucleus: Atoms, Molecules, Ions and the Chemical Bond

Atoms, molecules and ions represent the construction bricks of everything around us, including our body. The concepts of atoms, molecules, ions and the chemical bond must constitute the point of departure in studying chemistry. The relationship between chemistry and language permits a comparison of the atoms with the letters of the alphabet, and the molecules with the words. Just as the words serve to construct sentences that express ideas, so too are molecules the fundamental components whose combination gives rise to the complex world that is in us and around us.

> **The bricks and their interactions**

The description of how the atom, whose existence is now well accepted, and no longer needs to be verified, is made, must be simplified greatly. The principal objective here is to demonstrate that there is a recurring trend in the distribution of the electrons around the nucleus. This permits the ordering of the atoms into groups and periods, and leads to the construction of the Periodic Table (Periodic System). We are convinced that the time dedicated to the analysis of the Periodic Table is never too much. In this regard, it is good to recall that the Periodic Table is a single and concise document that summarizes a substantial part of the chemical knowledge—no other scientific discipline can claim to have a similar document.

Formation of a chemical bond and the type of bond can be interpreted on the basis of (a) the tendency of the atoms to share their odd electrons, and (b) to the different forces with which the various atoms attract the electrons involved in the interaction. It is also essential to explain that atoms, ions and molecules are *objects* so small that their dimensions can be guessed only through suitable comparison with objects we see and touch daily. On these bases, it is easy to insert the concept of *mole* (mol) to express, using reasonable numbers, the enormous quantity of atoms, molecules and ions present in samples of macroscopic substances.

Structural formulas and molecular geometry are clearly of fundamental importance in the interpretation of chemical properties. To simplify and make this part more interesting, the use of molecular models is recommended, including visual computer models. More time should be dedicated to describing some highly important molecules that are relevant in daily activities.

Some details on how the concept of the atom and, most importantly, how the concept of the molecule evolved historically should be taught, so as to make the students understand how much labor it took to have the existence of such invisible and mysterious entities accepted by the scientific community, and by the chemists themselves.

8.2.2 Second Thematic Nucleus: The States of Aggregation of Matter and the Solutions

| Solid, liquid and gas |

The first concept to discuss is that the normal state of aggregation of a chemical species at ambient temperature and atmospheric pressure depends on the type of bonding between the atoms. If the bonding involves a large number (theoretically infinite) of atoms (*e.g.* diamond or some other metal) or ions (*e.g.* sodium chloride), the species is a solid. If, on the other hand, it involves a defined number of atoms, with the consequent formation of molecules, the *state of aggregation* under normal conditions could be solid, liquid or gas. Normally, small molecules interact little with one another and thus are likely to be gases, whereas larger molecules have a greater tendency to interact, thereby giving rise to either liquid or solid species. Among the various types of interactions between molecules, one should describe and underline the importance of the hydrogen bond that is responsible for the anomalous properties of water, which are essential for the existence of life on Earth. Other fundamental concepts are the conditions of temperature and pressure, which determine the changes from one state of aggregation to another, a characteristic of every chemical species that could be used to identify the substance and its purity.

> **It dissolves or it doesn't dissolve?**

Solutions should be described as a common occurrence, since a large number of reactions take place in solution, especially those of biological importance: for instance, sea water is a solution, the processes that regulate life occur in solution and air is a solution (or mixture) of gases. It should be emphasized that gaseous, liquid and solid solutions have in common a disordered distribution of the species, but they are profoundly diverse with regard to interactions at the atomic/ionic/molecular levels. Speaking of solutions, particular attention should be dedicated to the colligative properties that have implications as much in daily life as in the biological world. In addition, solutions lend themselves to interesting experimental displays.

8.2.3 Third Thematic Nucleus: The Chemical Reactions

Without doubt, this thematic nucleus is the most important, as it constitutes the pivotal issue in chemistry. Consequently, both attention and time need to be dedicated to this topic.

To facilitate an understanding of a chemical reaction, it is wise to refer to events that occur continuously around us, recalling that even conception and death are the results of complex sequences of chemical reactions.

Oxidation–reduction reactions should be discussed in as simple a manner as possible limiting, to the maximal extent, the theoretical part so as to emphasize their relevance in the natural world and in the various fields of human activities.

Transformations of chemical species

Insofar as acid–base reactions are concerned, the first concepts to tackle regard the definitions of these substances that prescind from various theories. They must be simple and of immediate understanding. Even in this case, much attention should be dedicated to show the importance that such reactions have in the environmental, industrial and biological fields.

With regard to complexation reactions, it is important to emphasize that many important molecules in life, such as hemoglobin and vitamin B12, for example, contain metal ions. As well, chemical reactions that are accompanied by spectacular manifestations (gas evolution, precipitates and color changes) lend themselves nicely to class demonstrations. They can stimulate the students' interest and confirm that chemistry is, after all, an experimental science.

The concept of spontaneity of a chemical reaction must certainly be discussed, albeit simply as a balance between two factors that may cooperate or otherwise oppose each other: heat exchange (energy) against the variation in the level of disorder (entropy). It is also appropriate to underline the role played by other external factors (for instance, light) that can intervene and make a process spontaneous, even though the reaction may not be. It must also be remembered that, among other factors, light induces chemical reactions that allow life to exist on Earth.

Chemical equations must be introduced in terms of a symbolic and synthetic language adopted by chemists to describe a process, pointing to the fact that they provide qualitative information through the chemical formulas of the reagents and products, and quantitative information through their respective stoichiometric coefficients. Some numerical exercises would be useful to emphasize the significance of balancing chemical equations so as to highlight the importance of the *mole* concept, a convenient means between the microscopic world and the macroscopic world, and to demonstrate that through stoichiometry it is possible to predict the quantity of substances consumed or produced by a reaction.

Within this context, it is particularly important to tackle the concept of the *limiting reagent*, that is, the reagent present in least quantity relative to what is required by the stoichiometry of the reaction. Using simple calculations and examples taken from daily activities—for example, how many pies can be made with a certain quantity of ingredients—it is possible to demonstrate that in all chemical reactions (even those that occur in our body) the quantity of products formed is determined by the limiting reagent.

Moreover, the study of reactions must also take the time factor into consideration, a parameter not included in the concept of spontaneity. Among other aspects to tackle, the possibility of influencing the rate of a spontaneous reaction through variations in temperature and utilization of catalysts and inhibitors merits particular attention.

When the nuclei are transformed

Finally, the Chernobyl nuclear disaster and the more recent Fukushima nuclear incident are topics of a long-standing debate on the use of nuclear energy and the ever-increasing use of gamma radiation and other high-energy radiations. For students to understand *nuclear transformations (nuclear reactions)* requires that they be provided with the appropriate information. Once again, explanations should be limited to a few concepts, which point out the notion that reactions between nuclei cause the atoms to be transformed into atoms of different elements. As they do so, they release a considerable amount of energy, far greater than the quantity involved in normal chemical processes for which only the (peripheral) electrons of the atoms are involved.

It is important to have students understand that radioactive substances must be handled with extreme caution because of the obvious risks they imply, without demonizing all that is related (even if far related) to the nuclear question. The use of nuclear reactions for warlike purposes must certainly be deplored, and it should also be pointed out that nuclear power plants are not the right answer to energy issues for several reasons, including the risks they entail. However, it must not be forgotten that radioactive substances, and the accompanying radiations emitted upon their decay, do have certain useful and beneficial applications: for example, in the sterilization of medical products, in polymerization processes, in conferring special properties to materials of various types, in medical diagnostics, in radiation therapy against tumors, and in dating archeological artifacts, among others.

8.2.4 Fourth Thematic Nucleus: Chemistry in Everyday Life

> **Chemistry: a pivotal science**

A chemistry course that is exhaustive in the clear and correct description of the elements, molecules, bonding and chemical reactions, but which fails to examine the problems that one encounters daily, would miss its most important educational objective. It's important that the elementary language of chemistry be used to deepen the understanding of Nature and to point out the direct implication of this discipline for the bigger questions that society faces: food, water, energy, health, environment and information. By their very interdisciplinary nature, the arguments to tackle are those that lend themselves to lectures in the presence of professors from other disciplines so that these questions can be debated. For instance, in a science and society module, one could discuss the various ethical, cultural and social implications tethered to these themes by confronting the opinions of science professors with professors of the arts, philosophy and religion.

8.2.5 Fifth Thematic Nucleus: Chemistry Toward the Future

After having discussed the basic aspects of chemistry and realizing the importance of this discipline in understanding Nature and in resolving the problems of society, it would be useful and thought-provoking to look to the future to complete the training of young people.

> **A look to the future**

For instance, there are great expectations of nanotechnology, a technology that deals with matter at the nanometric level (a billionth of a meter) and thus at the molecular level. The problems of miniaturization require a totally different, yet novel, approach. The development of more powerful and fast computers and more sophisticated software will make it possible to simulate, with far greater precision, the course of chemical reactions, with a concomitant reduction of the times needed to perform the experimental procedures.

The language and methods of chemistry open up new horizons in biology and medicine, so much so that it doesn't seem so far-fetched that, in the future, machines at the molecular level that can intervene in single cells will be fabricated. It is also certain that chemistry will play a significant role in the quality control of products and in the monitoring of work environments. Indeed, any modern nation needs to possess companies staffed with well-trained, competent, capable and (why not?) creative chemists.

8.3 HOW TO TEACH

Laboratory teaching and the merry-go-round of curiosity

The study of chemistry from a textbook may seem interesting to a motivated student. However, to see chemistry *in action* and have *hands-on experience* through appropriate experiments would certainly be much more fascinating and stimulating, particularly to lesser motivated students. Using this teaching approach, commonly called *laboratory education*, allows the student to enter the true world of chemistry and appreciate its appeal and the satisfaction that comes with doing research.

Laboratory activities are in themselves scientific research for those that have yet to acquire a basic knowledge. It puts the student into that marvelous merry-go-round (Figure 6.2) that begins to move by the curiosity it engenders, and that feeds itself by questions that will find answers through suitably designed experiments.

Do and learn

The results of such experiments will generate knowledge as well as new questions that further stimulate curiosity. Then begins the second trip of the merry-go-round, at the end of which, enraptured by the fascination of discovery, no one wishes to descend.

Many pedagogues support the need for experimental activities that allow students to examine concrete facts, to talk about them, to explain the phenomena, and to classify the acquired knowledge. In this regard, the noted French pedagogue and educational reformer Celestin Freinet stated:

Knowledge does not happen through the study of rules and laws, but through experience. To study these rules and laws initially is like putting the cart before the horse. Rules and laws are either the result of experiences or else are worthless formulas.

As further support of this view, the American educational psychologist and philosopher Jerome Bruner asserts that: *People learn by doing, not by watching or listening, and learn better when they wish to know or have the desire to know.*

Learn from mistakes

Laboratory work, then, constitutes not only a moment of direct observation, but also one of analysis, of challenges, of comparison and verification, of formulation, of interpretations and predictions, and of invention and activity. Therefore, the inquiry-based laboratory activities help the student to understand, stimulate thought, create concrete experiments, and promote an active and personal processing of knowledge. In other words, this kind of teaching/learning is based on an education that starts from one's own need to learn, that teaches cooperation to organize experiments, transforming them into know-how, and exploits flexible routes recognized by the student as being meaningful.

Attempt to give meaning to words

The hands-on inquiry-based method is not specific to scientific disciplines, but is an approach that, with well-established methods of research and problem solving, focuses on acquiring knowledge and proficiency, rather than a multitude of facts. The laboratory should not be taken simply as a closed space equipped with instruments for students to carry out this or that experiment and demonstration. Rather, it should be experienced as an environment that provides opportunities for students to observe, plan and experiment. Accordingly, it is a setting wherein know-how is acquired and from which all the formative cross-sections of an observational, logical and linguistic nature are gained. These can be used later to produce new knowledge and develop new expertise, regardless of learning styles. From this perspective then, the educational mission moves from instruction to learning; that is, to processes of learning and reflecting that make students aware of the processes being experienced.

To give one pause for thought is very important to avoid the student learning only the words, instead of assimilating concepts. In this regard, the American philosopher, psychologist and educational reformer John Dewey said it more appropriately: *Attempting to give meaning only through the word, without any relationship with the thing, means to deprive the word of every intelligible explanation.*

In a sense, the laboratory environment is not unlike the Renaissance workshop, wherein everything started by creative experimentation. The apprentice learned by doing and seeing things done, by communicating with his master and his peers, and by visually stealing that which later became a technique. Laboratory activities favor the learning process in the format *do and learn*, to which subtends a strong motivation to undertake building/rebuilding one's own model of reality and exploit mistakes made. The latter become an efficient means to bring one's own knowledge to maturity. As was elegantly stated by an old maxim—*one learns through mistakes.* This, however, is true only if the student is not forced to duplicate a procedure given by the instructor. Rather, he devises the experiment autonomously, so as to experience personally the pleasure of experimentation.

> **Compare yourself with others and learn**

Another particularly interesting aspect of this teaching approach regards the fact that laboratory activity is normally organized in groups, so that the learning experience is lived in a relational context. This means that *do and learn* is integrated with *compare yourself with others and learn*, which leads to the construction of models shared by others and to the exploration of facts. The laboratory is then also the place and the environment wherein social relationships mature, because during the cooperative work, one also learns communication skills, leadership, negotiated solutions to a problem, management of conflicts and, above all, problem solving.

> **The role of the teacher**

In such situations, students and teachers have well-defined roles that transform the guiding principles of traditional teaching methodologies and put the student-protagonist at the center of the relationship and at the center of the process of teaching-learning. The teacher is placed flat second in this context. He is the organizer, the guide and the facilitator of the learning process. The teacher is the *director*, who must know how to create the proper environment and prepare the teaching stage within which every student is invited to use all the resources of rationality, creativity and talent, exactly as researchers do when engaged in resolving complex problems.

Obviously, this does not mean holding a detached and passive attitude. On the contrary, the teacher must participate with delight in the discoveries of his students. He should acknowledge new ideas with enthusiasm, ideas that could reveal themselves interesting and innovative. As was justly said by the Roman poet and philosopher Seneca (Lucius Annaeus Seneca; ca. 4 BC to 65 AD): *There is a twofold advantage in teaching, because, while teaching, one learns.*

The great potential of the experimental approach has been described nicely in an article that appeared in the American chemical literature, in which the central character is a young man, Ira Remsen, who subsequently became a famous chemist and the co-discoverer of the sweetener, saccharin. The story goes something like this:

| **Those remarkable words: *act on*** |

While reading a textbook of chemistry I came upon the statement, nitric acid acts upon copper. I was getting tired of reading such absurd stuff and I was determined to see what this meant. Copper was more or less familiar to me, for copper cents were then in use. I had seen a bottle marked nitric acid on a table in the doctor's office where I was then doing time. I did not know its peculiarities, but the spirit of adventure was upon me. Having nitric acid and copper, I had only to learn what the words act upon meant. The statement nitric acid acts upon copper would be something more than mere words. All was still. In the interest of knowledge I was even willing to sacrifice one of the few copper cents then in my possession. I put one of them on the table, opened the bottle marked nitric acid, poured some of the liquid on the copper and prepared to make an observation. But what was this wonderful thing which I beheld? The cent was already changed and it was no small change either. A green-blue liquid foamed and fumed over the cent and over the table. The air in the neighborhood of the performance became colored dark red. A great colored cloud arose. This was disagreeable and suffocating. How should I stop this? I tried to get rid of the objectionable mess by picking it up and throwing it out of the window. I learned another fact. Nitric acid not only acts upon copper, but it acts upon fingers. The pain led to another unpremeditated experiment. I drew my fingers across my trousers and another fact was discovered. Nitric acid acts upon trousers. Taking everything into consideration, that was the most impressive experiment and probably the most costly experiment I have ever performed ... It was a revelation to me. It resulted in a desire on my part to learn more about that remarkable kind of action. Plainly, the only way to learn about it was to see its results, to experiment, to work in a laboratory.

F. H. Getman, "The Life of Ira Remsen,"
Journal of Chemical Education, 1940, pp. 9–10.

When the spark is triggered by curiosity, the study loses much of its connotative obligation. It becomes a desirable occasion to learn to know. Albert Einstein, who was very conscious of the incredible importance of curiosity, would certainly agree. He said of himself: *I have no particular talents, I am only passionately curious.*

CHAPTER 9

Today's Science: Objectives, Implications and Limits

9.1 THE DEVELOPMENT OF SCIENCE

Scientific knowledge was rather limited at the time that the oldest athenaeum of the western world was founded in 1088: the University of Bologna, Italy.

> **Nearly nothing of nearly all; nearly all of nearly nothing**

One could say that, in those old days, scientists knew *nearly nothing of nearly all*. Indeed, in their attempt to understand the laws of Nature, they limited themselves to contemplating Nature, thereby remaining at a somewhat superficial level of understanding. A few centuries later, around the 1600s, scientists like Galileo and Newton recognized that more could be done. Other than observing the various natural phenomena, these two scientific giants came to realize that Nature could be interrogated through experiments, thereby forcing her to reveal her secrets. Thus began the impressive development of science. With time, fields of knowledge broadened and ultimately fragmented into various disciplines.

To discover or to invent something new today requires serious research efforts in extremely specific fields. To be highly

---

Chemistry: Reading and Writing the Book of Nature
By Vincenzo Balzani and Margherita Venturi
Translation by Nick Serpone
© The Royal Society of Chemistry 2014
Published by the Royal Society of Chemistry, www.rsc.org

specialized has become a necessity. Therefore, contrary to what occurred centuries ago, today's scientists know *nearly all of nearly nothing*. And when they speak of that nearly nothing that they know so well, scientists often tend to use words that no one can understand.

> By nature scientists have the future in their blood

The expansion of knowledge has had many obvious positive consequences, but also its problems. One of these is the unavoidable fracture created between the various branches of science, as every discipline has developed its own language. The indispensable collaboration between scientists of various disciplines is therefore too often difficult.

The precarious break between the sciences and the humanities was first announced about fifty years ago by Charles P. Snow, who pointed out: *The humanists have their eyes focused on the past, while (by nature) scientists have the future in their blood*.

It is imperative that such a break be mended as only with knowledge of the past can we live well in the present and build a better future for the next generations. The humanities must be able to communicate with the physical sciences because the complexity of the world requires that the various disciplines interact with each other. Progress can only occur through discussions among diversities.

During the last few years, an even more dangerous break has taken place, that between science and society. The scientist of today is undoubtedly in a difficult situation. On the one hand, he has to master his field, he must be specialized, and at the same time he must maintain contacts with other disciplines. On the other hand, if he wants to be a good citizen he must also be interested in issues that concern his community, and must publically assume those responsibilities that come with the knowledge acquired.

9.2 WILL SCIENCE COME TO AN END?

Articles and books appear periodically that, while emphasizing the great development of science during the last few years, maintain that by now all or nearly all has been discovered. Science is destined to end soon. It has no future!

This is also the opinion of many who believe that all that happens in the world can be reduced to properties of elementary particles, framed in theories expressed by mathematical equations. For instance, in the book titled, *The Theory of Everything*, the renowned English astrophysicist, Stephen Hawking, maintains that:

> *if we were intelligent enough to discover this unified theory, we would decree the definitive triumph of the human spirit, because then we would know the spirit of God.*

If young people intend to pursue a career in a scientific field, they needn't worry. Science is not about to come to an end any time soon. In fact, it is just the beginning! Science is still in its infancy!

> **The larger the circle of light is, the greater is the margin of darkness.**

Some years ago, the magazine *Science* thought of writing a directory of the 25 most important questions that have yet to be answered by scientific research. In their attempt to carry out this task, the journal editors were at first forced to extend the directory to 125 questions. Soon thereafter, however, they surrendered, as they realized that there were still far too many other questions because:

> *The freeway that goes from ignorance to knowledge extends in both directions: as knowledge accumulates, ignorance of the past diminishes, but new questions arise that expand the area of ignorance remaining yet to be explored.*

In fact, beyond the many things that we know we don't know, there are also things we don't know we don't know. The concept that science will never end was expressed poetically by the 18th-century English theologian, educator, natural philosopher and chemist Joseph Priestley, who was also the first scientist to study photosynthesis. He stated:

> *The greater is the circle of light, the greater is the boundary of the darkness by which it is confined. But notwithstanding this, the more light we get, the more thankful we ought to be. In time the bounds of light will be still farther extended; and from the infinity of the divine nature, and the divine works, we may*

promise ourselves an endless progress in our investigation of them: a prospect truly sublime and glorious.

Consequently, a scientist can never claim to *know it all*, not even in his specific field of research. As pointed out by the philosopher Martin Buber, *if you have gained knowledge, then only you know what else you are missing.*

9.3 CHARACTERISTICS AND LIMITATIONS OF SCIENCE

Science is important because it tells us how the world is made, how man is made and how we can change both. Thus, science takes on a great responsibility for better or for worse. Science is useful because it teaches us how to fight diseases, how to avoid troubles, how to better enjoy life and the many other advantages that it entails. We must not forget, however, that when used poorly, science can be terribly dangerous. Science also is fascinating because to discover the mysteries of Nature—from the infinitely large to the infinitely small realities—can only cause astonishment and surprise.

Science is not only important because of the many material benefits it provides, but also because it educates people about democracy. In fact, the characteristics of science (exactitude, objectivity, doubt, comparison, collaboration, freedom of thought, acceptance of dissent, refusal to submit to impositions) are the same as the pillars of democracy. Scientific knowledge is collective knowledge, a sort of a big building, constructed stone by stone by a large number of people: architects, engineers, specialized workers and laborers. To do so, more sophisticated instruments are used in always increasing number. Sometimes it happens that a portion of the building may need to be demolished because it was built with a badly conceived plan, or else with poor materials. In this regard, Richard P. Feynman stated:

Science is the results of the discovery that it is worthwhile re-checking by new direct experience, and not necessarily trusting the experience from the past.

| We are like dwarfs on the shoulders of giants |

Knowledge accumulated through time is of utmost importance and valuable to new investigators, as they push the frontiers of scientific research even further using procedures established by their predecessors. As noted by Bernard of Chartres and, later on, by Newton, *If I have seen further it is by standing on the shoulders of giants.*

> **Science is based on the exchange of ideas**

The time is no longer that of the scientist working alone in his ivory tower. The high degree of specialization of investigators and the necessity for an interdisciplinary approach to confront the more important issues require that research be performed by national and international groups through collaborations on projects such as those financed by the European Union. The exchange of ideas and notions play a dominant role in any collaboration. As the Irish playwright George Bernard Shaw noted in one of his famous quotes:

> *If you have an apple and I have an apple and we exchange apples between us, then you and I have but one apple each. But if you have an idea and I have an idea and we exchange these ideas between us, then we both have two ideas.*

(This quote has recently become popular in the less poetic version of the American author Dan Zadra who replaced the apple with the dollar.)

Science is important, useful and appealing, but it also has restrictions and limits. Science explains *how* but not *why* natural phenomena occur. For instance, we know that light has a twofold characteristic—that is, it acts as a particle and as a wave—and yet we don't know why. Likewise, science cannot provide answers to some very fundamental questions, the common-sense questions that emerge in every being: what sense does my life have? Why is there the mystery of evil? Does God exist? Answers to these questions must be sought elsewhere, in the fields of philosophy and religion. By itself, science cannot make us live in a more just world. This responsibility falls in the realm of politics. As the Greek philosopher Plato (427–357 BC) once said:

> *(Politics) must assert what is right through coordination and the government of all knowledge, the techniques and the activities that take place in the city.*

9.4 THE ROLE OF SCIENCE IN A FRAGILE WORLD

> It is the overwhelming power that we gave ourselves to impose us to recognize what we're doing

As science expands, technology develops. Scientific discoveries and inventions allow us to understand how the world functions and offer us instruments to change the world. This can be done for noble or for immoral purposes. Accordingly, this leads to ethical problems of a different nature and importance. In general terms, the more science is developed, the easier it will be to perpetrate malicious actions or make disastrous errors. Therefore, if it is true that results from scientific research and the consequent developments of technology can prove very useful, it is also undeniably true that such results make the world more complex and more fragile. The progress of science should then push us to act in a more mindful and responsible manner. As the philosopher Hans Jonas said:

It is the overwhelming power that we gave ourselves, on us and on the environment; it is the immense causal dimensions of this power that imposes us to recognize what we are doing, and to choose in which direction we want to advance.

Unfortunately, the choices we make are not always the right ones.

With the proliferation of increasingly sophisticated weapons and other detrimental applications of scientific knowledge, there may be a need to place certain limits on scientific research. This is a very complicated issue and a much debated one at that. Is it right? Is it possible? Since the objective of scientific research is to discover all truths (that is, *knowledge*) it does not seem reasonable to place limits on knowledge. Scientific research must be free. Indeed it is everyone's right. Perhaps we should put some limitations on its applications, and on research techniques. The distinction between pure and applied research, however, is anything but clear. We mustn't forget that to know and to act are doubly interlaced, because to act we need to know and to know we need to act. Just like everyone else, when the scientist acts, he does so on the basis of some ultimate purpose and merit that, by definition, is never neutral. Whenever the scientist acts with no purpose and no personal integrity, research then poses a greater danger because it plays into the hands of whoever has the financial resources and power. All this tells us that scientific research is not and cannot be neutral, neither in the choice of priorities, nor in the way in which it is carried out, nor much less in its effects on society. Hence, every democratic country must question itself and must make decisions as to the direction scientific research should take.

Priority should be given to investigations aimed at resolving the more important and urgent issues that plague our planet. In the specific case of chemistry, scientific research should be largely oriented in finding solutions to the energy crisis because the availability of food and water and a safer environment depend on the quality and quantity of usable energy.

9.5 THE SOCIAL RESPONSIBILITY OF THE SCIENTIST

> To look far: into the world, into the future

Because of the great social importance of his work, the scientist cannot be pleased just to be a master in his narrow field of study. He cannot close himself in his laboratory, fascinated by the beauty of his research and gratified by the accolades from his peers. Whoever works in scientific research must assume the responsibilities that come from the privilege of possessing knowledge and enjoying a position of respect in society. According to the 1991 Nobel Prize winner for Chemistry, Richard R. Ernst, scientists have a responsibility to establish the guidelines of real progress for humanity, to participate actively in social life, so as to inform people of the benefits and risks of science, and to favor the creation, with authoritative advice, of new policies that are forward-looking: far into the world for the good of all people on Earth, and far into the future for the good of the next generations.

Man comes before science! As justly stated by Albert Einstein:

Concern for man himself and his fate must always constitute the chief objective of all technological endeavors; never forget this in the midst of your diagrams and equations.

Chemists must remember this when they are immersed in the wonderful world of molecules.

It is necessary, therefore, to further the progress of science, but it is even more important to choose research priorities. It is important to use technology, but it is still more important to know the purpose for which it will be used:

- seek peace and not wage war;
- alleviate poverty and not maintain privileges;
- reduce and not increase the inequalities between nations;
- safeguard and not destroy planet Earth.

An African proverb says: *Planet Earth was given to us on loan by the next generations.*

Subject Index

acetic acid ($C_2H_4O_2$) 36
acetylsalicyclic acid (aspirin) 73
acid rain 44
acid-base reactions 44–5, 84
Act on term 104–5
adenine (A) 58–9
air bags 51
alanine (isomers) 33–4
ammonia (NH_3) 25–6, 29, 30–2, 41
ammonium chloride (NH_4Cl) 46
anesthetics (no pain) 78
anthracene 87
anti-oxidants 52
architecture and chemistry 80–1
Aristotle 19
aromatic molecules 17
artificial molecules 74–7
artificial products 6
aspartame 75
Aspirin 72–4
association of molecules 53–4, 56
atom
    concept 10
    term (*atomos, not divisible*) 13–14

atoms, spatial disposition 32
atoms and molecules -
    language of chemistry
    comparing matter with language 15–18
    history
        from Dalton to Cannizaro 20–2
        from Greek philosophers to Lucretius 19–20
        last 150 years 22–3
    perfume of roses 9–10
    what is the world made of? 10–15
Avogadro, Amedeo 21–2

balancing reactions 46–7
barbital 55
barium sulfate ($BaSO_4$) 40
batteries 42
Bayer 73
beautiful molecules 80–2
Beethoven 64
benzene (C_6H_6) 36
Bernard of Chartres 110
Berzelius 21–2
bioethics 63
biology and chemistry 7, 101

blue color (cornflower) 37
Bok, Derek 3
bond strength 26
Bruner, Jerome 101
Buber, Martin 109
Buchner, Johann Andreas 72
di-esa-tert-
 butyldecacyclene 28

caffeine ($C_8H_{10}O_2N_4$) 36
calcium carbonate
 ($CaCO_3$) 41, 83
calcium oxide 24
capsaicin (spice) 71
carbohydrates 58
carbon 25
carbon atoms 14
carbon dioxide (CO_2) 18, 26,
 41, 43, 47, 83
carbon monoxide (CO)
 29, 52, 71
carbonic anhydride *see* carbon
 dioxide
catalysts
 enzymes 52
 human body 52
 pharmaceutical
  industry 52
catalytic converters (cars) 52
cells 57, 60-1, 61-3
chemical bonds 24-7
chemical equations 46-8, 98
chemical industry 2
Chemical Industry
 Association 6
chemistry in action (reactions)
 acid-base reactions
  44-46
 equations 46-48
 oxidation-reduction
  (redox) reactions 42-43

reactions and time 50-52
 transformation of
  chemical species 39-41
 why reactions occur
  48-50
chemistry of life 62
chemophobia 6
Chernobyl nuclear disaster 98
chiral isomers 33, 52
cholesterol ($C_{27}H_{46}O$) 36
clam shells 40
combinatorial chemistry 74
combustion reactions 42-3
computational chemistry 74
conception of life (birth) 39
"cork" flavor 17-18
coupling between atoms 24-5
creativity and beauty
 beautiful molecules 80-1
 for better or for
  worse 78-80
 chemistry in words of
  scientists/writers 81-4
 intelligent molecules
  84-7
 molecular machines
  88-90
curiosity merry-go-round 70-1,
 101
cyanuric acid 26
cyclamate 75
cyclamen molecules 9-10
cyclictetramethylene-
 trinitroamine (HDX) 79
cyclictrimethylene-
 trinitroamine (RDX) 79
cytosin 55
cytosine (C) 58-9

da Vinci, Leonardo 74
Dalton, John 20-2

De rerum natura (philosophical poem) 16
deoxyribonucleic acid (DNA) 57–9, 62
Dewey, John 103
dimensions of molecules 27–9
dimethyl ether (ethanol isomer) 32
dioxin 77
disorder (entropy) 98
DNA polymerase 59–60
do and learn format (teaching) 103
double bonds 26
drugs
    diseases 78
    see also Aspirin
dyes 78
dynamite 79

education is not filling a bucket, but lighting a fire 95
Einstein, Albert 105, 112
electricity 90
electrochemical reactions 84
electron transfer 42
element concept 10
elements
    Earth 11
    human body 11
    universe 11
Empedocles 19
energy and disorder 48
enzymes
    catalysts 52
    DNA 59–60, 62
Epicurus 19
Ernst, Richard R. 112

essential science
    education science and culture 1–5
    image of chemistry 5–6
    importance of chemistry 6–7
    sciences and chemistry 7–8
estradiol ($C_{19}H_{24}O_2$) 37–8
ethane 26
ethanol (C_2H_6O) 30–1, 32
ethyl alcohol *see* ethanol
European Union (EU) and research projects 110
evaporation of water 49
explosives 79
eye and light 37

Faraday, Michael 40–1, 90
fertilizers 78
Feynman, Richard P. 69, 81, 109
flame of a lit candle 40
Flint, Frank S. 68
fluorescence 87
food and drink additives 78
Food and Drug Administration (FDA), US 73
formulas of molecules 30–4
Freinet, Celestin 102
Fukushima nuclear accident 98
fungicides 78

Galileo 106
Gay-Lussac 20–1
genetic engineering 62–3
genetically modified organisms (GMOs) 62
Gladstone, William 90

Subject Index

glucose ($C_6H_{12}O_6$) 15, 29, 71
Goethe 28
grapefruit flavor 17–18
Gross Domestic Product (GDP)
   in United States 2
guanine (G) 55, 58–9

hands on experience 100
heat exchange (energy) 98
heat (thermal energy) 49
hemoglobin 18, 29
Hoffman, Roald 83–4, 88
host-guest system 55–6
hydrochloric acid (HCl) 30
hydrogen 25–6
hydronium (H_3O^+) ion 45

ICMESA Company, Italy 77
image (chemistry) 5–6
importance (chemistry) 6–7
from inanimate matter to
   life 57
incendiary bombs 79
increase of disorder 49
inhibitors 51–2
insect bites (repellants) 79
insulin ($C_{254}H_{377}N_{65}O_{76}S_6$) 71
intelligent molecules 84–7
isomerism 32
isomerization reactions 84
isomers 32

Jonas, Hans 111

knowledge 111
Kolbe (German chemist) 72

laboratory education 101
Lavoisier 20
lectin 61

Lego 35, 75
Lehn, Jean-Marie 88
Leucippus 19
Levi, Primo 13, 27, 80, 82–3, 88
life is chemistry in action! 6
life is more than chemistry and
   biology 64
light (photochemical
   reactions) 84
limiting reagent concept 98
lipids 58
lock-key model 55–6
logic gates
   AND 86–7
   OR 86
   XOR 86

manganate ion (MnO_4^{2-}) 45
manganese dioxide (MnO_2) 45
manganese ion (Mn^{2+}) 45
materials chemistry 74
memory and molecules 85
Mendeleev, Dmitri I. 13, 20
methane (CH_4) 25–6, 30–2, 46–7
minimal life 57
models of molecules 34–7
mole concept 96–7
molecular association 53–4, 56
molecular recognition 54, 60
molecule concept 27
molecules
   anesthetics 78
   artificial 74–7
   beautiful 80–2
   chemical complexity
      53–4
   description 14–15
   drugs 78
   dyes 78
   fertilizers 78

molecules (*continued*)
    food and drink
        additives 78
    fungicides 78
    information content 53
    insect bites 79
    intelligent 84–7
    logic gates 86–7
    memory 85
    nature 56, 67–8, 71–2
    perfumes 79
    photochromics 78
    shapes/properties 8
    sunscreens 79
    thermal insulators 78
    words of matter 16–18
    see also atoms and molecules; beyond molecules - from chemistry to biology; reading and writing with molecules; world of molecules
beyond molecules - from chemistry to biology
    from cells to man 61
    genetic engineering 62–3
    from molecules to supramolecular systems 53–6
    beyond scale of complexity 63–4
    from supramolecular systems to cells 57–61
Mondrian, Piet 69

names of molecules 29–30
nanotechnology 100
natural molecules 56, 67–8, 71–2

natural versus artificial products 67
nearly all of nearly nothing 107
nearly nothing of nearly all 106
Newton, Isaac 106, 110
nitrogen (N_2) 25–7, 47, 51
nitrogen oxides (NOx) compounds 52
nitroglycerin (NG) 79
Nobel, Alfred 79
nuclear transformations (nuclear reactions) 98

one learns through mistakes 103
optical isomers see chiral isomers
Oxidation-reduction (redox) reactions 42–3, 84
oxygen (O_2) 25–6, 28, 71

peperoncino (hot pepper) 71
perfumes
    creativity 79
    cyclamen 9–10
    roses 9–10
Periodic Table (System) 11–13, 16, 20, 96
permanganate ion (MnO_4^-) 44–5
pH
    color 45
    concept 45
    indicators 45–6
    meters 45
    skin 45
photochromics (eye protection) 78
photosynthesis 6
"pila" (battery) 42–3
Piria, Raffaela 72

Subject Index

Planet Earth was given to us on loan by the next generations 113
Plato 110
Priestley, Joseph 108
precipitation reactions 40
profession of chemistry 81-2
proteins 58-9
proton transfer 44

rate of reactions 51
reactions
    acid-base 44-5, 84
    balancing 46-7
    chemical equations 46-8
    combustion 42-3
    electrochemical 84
    isomerization 84
    nuclear 98
    oxidation-reduction 42-3, 84
    rate 51
    temperature 50-1
    time 50-2
    transformation of chemical species 39-41
    why they occur 48-50
reading and writing with molecules
    artificial molecules 74-7
    Aspirin 72-4
    chemist - explorer and inventor 67-8
    merry-go-round of curiosity 68-71
    natural molecules 71-2
recombinant DNA 62
red color (poppy) 37

redox reactions *see* oxidation-reduction reactions
release of heat 49
Remsen, Ira 104
respiration 43
11-cis-retinal $C_{28}H_{28}O$ 38
trans-retinal $C_{28}H_{28}O$ 38
ribonucleic acid (RNA) 58
rose molecules 9-10
rotaxane 89

saccharin 75
salicin 72
salicylic acid 72
salt precipitation 40
Science 63, 94, 108
sciences and chemistry 7-8
Seneca (philosopher) 104
Shakespeare, William 18
Shaw, George Bernard 110
silver chloride (AgCl) 41
Snow, Charles P. 107
Sobrero, Ascanio 79
sodium bicarbonate ($NaHCO_3$) 44, 46
sodium chloride (NaCl) 24, 46
sodium hypochlorite (NACLO, bleach) 46
solutions 97
spatial disposition of atoms 32
spontaneity in chemical reactions 48, 50
spontaneity concept 98
stearine ($C_{57}H_{110}O_5$) 36
stereoisomerism 33
Stone, Edward 72
structural formulas 31, 34-5
sucronic acid 76

sucrose 75
sugars 61
Sun (source of energy) 50
sunscreens 79
supramolecular systems
    cells 57–61
    logic gates 87
    molecules 53–6
Sustainable Development 84
sweeteners 75–6
Szent-Gyorgi, Albert 70

teaching chemistry: what and how
    how to teach 101–5
    school subject 93–4
    what to teach 94–101
teaching chemistry: what and how (themes)
    aggregation of matter and solutions 96–7
    atoms, molecules, ions and chemical bonds 95–6
    chemical reactions 97–100
    chemistry in everyday life 100
    chemistry towards the future 100–1
temperature, reactions 50–1
termination of life (death) 39
testosterone ($C_{19}H_{28}O_2$) 37–8
textiles 8
thalidomide 33–4
The Monkey's Wrench 27, 82–3
The Periodic System 80
The Red Tree 69
Theophrastus (Greek philosopher) 95

thermal insulators 78
thymine (T) 58–9
time and reactions 50–2
Titus Lucretius Carus 16
today's science: objectives, implications and limits
    characteristics/limitations 109–10
    development 106–7
    role in fragile world 111–12
    social responsibility 112–13
    will science come to an end? 107–9
Tolstoy, Lev N. 69
toluene (methylbenzene) 75
tomato sauce 44
transgenic organisms *see* genetically modified organisms
Trees (poem) 68
trinitrotoluene (TNT) 79
triple bonds 26–7

United States (US)
    FDA 73
    Gross Domestic Product 2
universal catalyst 51

vitamins
    B12 98
    C 52
Volta, Alessandro 42–3

War and Peace 69
water (H_2O)
    bonds 25–6
    cells 58

chemical equations 46
dimensions 28
elements 15
evaporation 49
formula 31
molecular dimensions 27
names of molecules 39
respiration 43
structural formula 30
words of matter 16–17

wine bouquet 17
words of matter 16–18
world of molecules
    attention to details! 37–8
    chemical bonds 24–7
    dimensions 27–9
    formulas 30–4
    models 34–7
    names 29–30

Zadra, Dan 110